T0362729

Robert Vennell is the bestselling author of *The Meaning of Trees: The history and use of New Zealand's native plants*. He works in the Natural Science department of Auckland War Memorial Museum – Tāmaki Paenga Hira, and currently lives with his wife in an ocean-lover's paradise in Ngunguru, Northland.

Secrets of the Sea

ROBERT VENNELL

HarperCollins*Publishers*
Australia • Brazil • Canada • France • Germany • Holland • India
Italy • Japan • Mexico • New Zealand • Poland • Spain • Sweden
Switzerland • United Kingdom • United States of America

First published in 2022
by HarperCollinsPublishers (New Zealand) Limited
Unit D1, 63 Apollo Drive, Rosedale, Auckland 0632, New Zealand
harpercollins.co.nz

A catalogue record for this book is available from the National Library of New Zealand.

ISBN 978 1 7755 4179 0 (hardback)

Cover design by Emily O'Neill
Internal design by Emily O'Neill, based on an original design by Hazel Lam,
HarperCollins Design Studio
Typeset in Baskerville by Emily O'Neill
Colour reproduction by Splitting Image Colour Studio, Clayton VIC
Printed and bound in China by 1010 Printing on 140gsm woodfree

6 5 4 3 2 1 22 23 24 25

For Dad

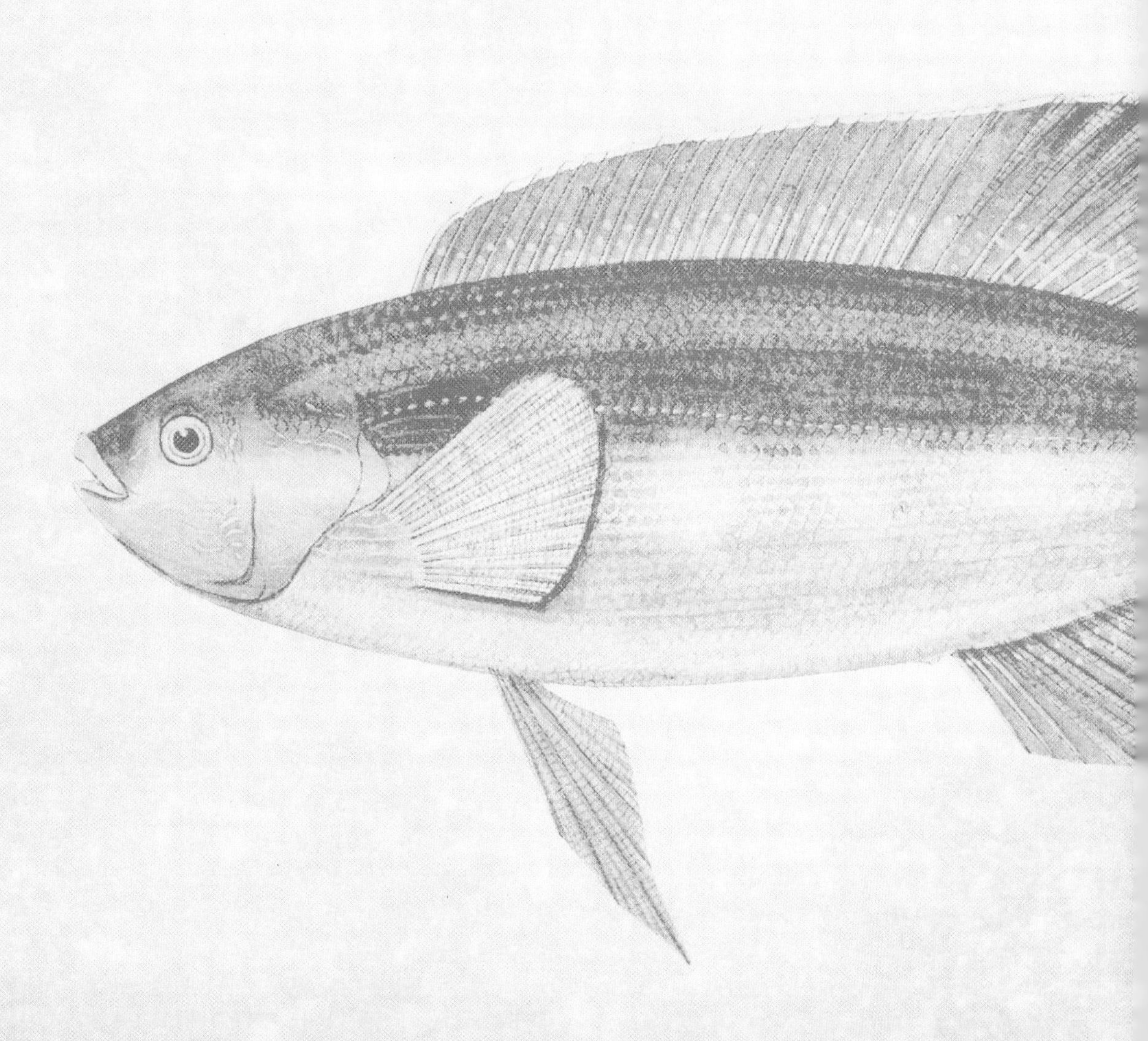

CONTENTS

5. Denizens of the Deep

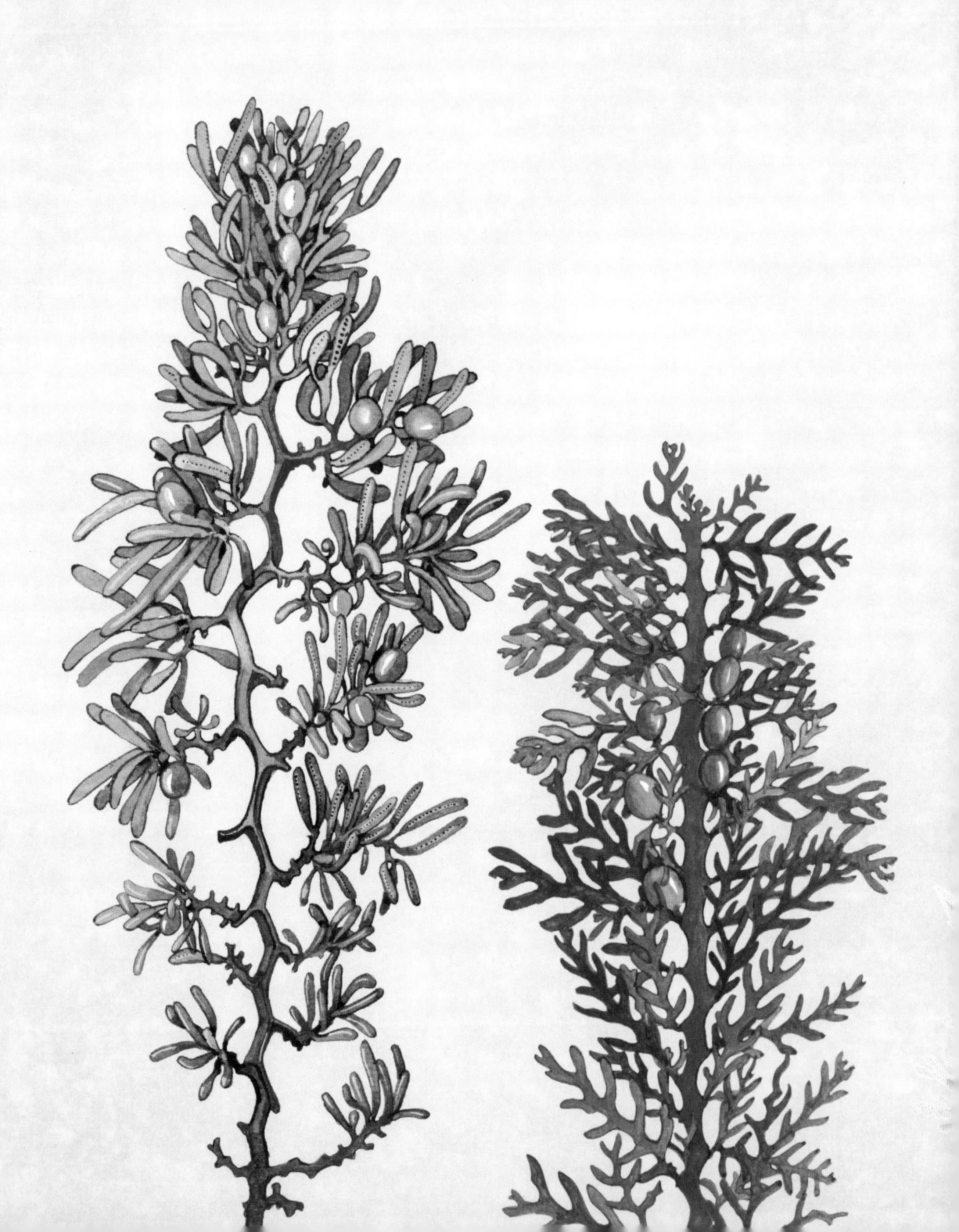

PREFACE

The oceans of Aotearoa are full of wonder, magic, mystery and awe. Hidden beneath the waves are some of the most fascinating and unique creatures in New Zealand, all with remarkable stories to tell. And yet, because they live secret, underwater lives they can be easy to overlook. As we go about our daily life on land, sea creatures can remain out of sight and out of mind.

With this book, I want to shine a light on our incredible native fish and shellfish, and show that, even though these creatures live in a different world from us, they have profoundly shaped our lives on land. Since the very first people arrived in Aotearoa, fish and shellfish have been critical to the human story. They have helped us to survive on this oceanic island in the middle of the South Pacific. They have been a way for us to connect with one another, to trade, to feast, to celebrate and to mourn. At times they have been hated, and even hunted to extinction. At other times, they have been regarded as sacred and have inspired great works of art, poetry and song. Some have languished in obscurity or been lost to time, while others have become central to a shared New Zealand identity.

Our oceans are under siege, and finding new ways to connect with our native sea creatures is more important than ever. I hope this book helps inspire a new appreciation of New Zealand's marine life by revealing some of the secrets of that mysterious world right on our doorstep.

Robert Vennell,
November 2021

HOW THIS BOOK IS ORGANISED

The book begins with a brief exploration of the role fish and shellfish have played in New Zealand history, before focusing on the creatures themselves and their individual stories. It is broken up into five chapters, which start on land then move further away from the coast and into the deeper reaches of the ocean. Some species could arguably have been placed in several different groups; in those cases I have chosen what I thought was the most natural fit. The five chapters are:

- Fresh Water: creatures of the rivers, streams and lakes
- Sandy Shores: creatures of estuaries and sandy beaches
- Rocky Reefs: creatures of the reefs and kelp forests
- Ocean Hunters: creatures that hunt other fish out at sea
- Denizens of the Deep: creatures from the ocean depths or that live far out to sea

When writing a book like this, one of the most difficult aspects is deciding what to include and what to leave out. Rather than attempt to comprehensively survey the ocean, my goal has been to present a smorgasbord of stories, including a sampling of different species that occur in New Zealand waters.

When choosing which creatures to focus on I aimed to include iconic species alongside some lesser known or forgotten ones. I wanted to keep the stories diverse and avoid repetition, which is why some well-known and beloved species missed out (e.g. pipi, oysters and green-lipped mussels). Although the book is primarily about sea creatures, I also couldn't help but include a few freshwater species that never travel to the sea, as they felt too important to leave out of the story.

A NOTE ON TAXONOMY

When you say the word 'fish', most people have a general idea of what you mean – a slippery creature with gills and scales that lives in water. But there's a lot more to the term than meets the eye. Fish are one of the largest groups of animals, with over 30,000 incredibly diverse species – more than all mammals, birds, reptiles and amphibians put together. Within the fish there is so much diversity that a bony fish like a snapper is more closely related to a human than it is to a shark.

Then there are the marine invertebrates, which make up the overwhelming majority of animal life in the ocean. All of the basic animal body plans are represented by the marine invertebrates, including creatures as diverse as jellyfish, worms, crabs, octopuses, mussels, starfish, corals and sponges. I've rather simplistically referred to these creatures as 'shellfish' in the text – although not all shellfish have shells and none of them are fish in the sense described above.

The science of taxonomy is full of these incredible revelations, but a little basic knowledge is required to understand them. All of life can be organised into a taxonomic family tree based on the following descending groups:

- Kingdom
- Phylum
- Class
- Order
- Family
- Genus
- Species

A helpful mnemonic for remembering this structure is: 'Kina Poke Children On Feet; Get Shoes!'

Where a living thing fits into this structure determines how it is related to everything else on Earth. For example, the taxonomy of the New Zealand longfin eel looks like this:

- Kingdom: Animalia
- Phylum: Chordata
- Class: Actinopterygii
- Order: Anguilliformes
- Family: Anguillidae
- Genus: *Anguilla*
- Species: *dieffenbachii*

These Latin terms can seem a little intimidating at first, but at a glance they let us know where longfin eels sit in the tree of life. They tell us that longfin eels belong in the same group as the other back-boned animals (Phylum: Chordata), that they are a bony fish (Class: Actinopterygii), and that they belong to the freshwater eel family (Family: Anguillidae). They also give us a completely unique name for the species, which distinguishes it from every other creature on earth: *Anguilla dieffenbachii*.

Another helpful distinction to keep in mind is the difference between an endemic or a native species. *Anguilla dieffenbachii* is 'endemic' to New Zealand, meaning it naturally occurs here and nowhere else, whereas the shortfin eel (*Anguilla australis*) is 'native' – it occurs here naturally but can be found in a number of other places across the South Pacific as well.

A NOTE ON FISH AS FOOD

This is a story about how fish and shellfish have influenced our lives and the way we think about the world. It's about science and conservation, but it's also about food. It might seem unusual for a book whose purpose is to highlight the beauty and value of our marine species to talk so much about killing and eating them. My personal view is that sea creatures bring joy and meaning to our lives and there is nothing inherently wrong with eating them, providing we give them the respect they deserve and know when we are free to take and when we need to leave them alone.

INTRODUCTION

A brief history of sea creatures

The first people to arrive in Aotearoa were truly tangata moana – people of the sea. For thousands of years, Polynesians had travelled across the waves, discovering all of the tiny islands in the vast Pacific Ocean. They had explored more of the planet than any other people before them, and they had done so without compasses, maps or written instructions. But around 750 years ago, a small group of Polynesian explorers did something completely different. They turned south, leaving behind the warm tropical seas and sailing into the unknown. There was no telling what they might find, so they used the sea as their guide, reading meaning from the waves and winds and deciphering instructions from the movements of birds, fish and whales. Finally, they spotted it, stretching out like a long white cloud on the horizon: Aotearoa – the last major landmass in the world to be discovered by people.

For these ocean explorers, their first introduction to Aotearoa would undoubtedly have been through its sea creatures. The oceans surrounding these islands were bursting with fish and shellfish, never before touched by human hands. Out at sea, there were unbroken streams of snapper stretching across the horizon, while kahawai leapt into the air as if the ocean had become too crowded for them. As the explorers ventured closer to land, coastlines sparkled with shiny green mussels and crayfish antlers stuck out of the water, waving at the explorers as they passed by. As they reached the shore and leapt off their waka in excitement, the sea floor would have erupted with life: flounder speeding away in every direction, startled scallops taking off like a flock of birds, and giant toheroa digging themselves into the sand with their tongues. In the crystal-clear estuaries there were vast cockle gardens patrolled by sentinels of stingray, and every river bend was guarded by monstrous eels, like tree trunks that had fallen into the water and come to life.

When these explorers returned home to the tropics, they told stories of the islands they had discovered. It wasn't long before more voyagers set out to make Aotearoa their home. At first, food was everywhere in this new land. Flightless birds made their nests on the ground, with no fear of people; massive sea lions lounged around the coastline; and giant moa stalked the forest. But, as time went on, many of the birds of the forest were hunted to extinction and the sea lions vanished from around the coast. With the disappearance of the major food sources on land, the sea became more important than ever.

But while the oceans contained an abundance of food, it wasn't enough to simply throw out a net and hope for the best. If people were to survive in Aotearoa, they would need to understand everything they could about the fish and shellfish that lived here. As the new arrivals carefully observed the natural world, the sea slowly began to reveal its secrets.

The Legend of the Voyage to New Zealand by Kennett Watkins. (Auckland Art Gallery Toi o Tāmaki, U/32)

TE AO MĀORI

As generations of people grew up in Aotearoa and adapted to the new environment, they developed a rich culture and worldview known as Te Ao Māori. Fish and shellfish were central to that shift, and helped to set the rhythms and rituals of Māori life.

When the autumn rains came, eels could be caught as they swam downstream, and when Matariki rose in the sky, lamprey were collected as they suckered themselves up rocky waterfalls. When the kōwhai sprang into golden blossoms, the roe of the kina would start to become sweet and rich, and when the pōhutukawa flowered, snapper would bite in huge numbers. Permanent settlements were based around abundant fish resources, such as the barracouta grounds of the South Island or the abundant eels and freshwater mussels of the central North Island. When the kahawai or whitebait began to run in the rivers, Māori would travel across country and camp in temporary settlements for several months of the year, catching, drying and preserving the fish.

From an early age, children were taught the knowledge they needed to enter the ocean realm and bring back kai and other resources. They were taught to weave nets of different sizes and shapes, and how to craft hooks from wood, shell and bone. They learned how to dive for kina and crayfish, dredge for mussels, and gather cockles and toheroa from the sand. As they grew older, they joined in the weaving of enormous communal nets. When finished, these nets could weigh several tonnes, and the strength of the whole community was needed to haul them in from the sea, harvesting tens of thousands of fish in a single sweep.

With such an abundance of delicious seafood, a rich culinary tradition developed. All parts of the fish were eaten, which meant Māori could experiment and learn the choicest cuts – the livers of sharks, the entrails of the marble fish, or the eyes of the hāpuku. Each marae had its own special delicacy and it was a great source of pride to be able to welcome guests for a feast and serve them the local dish. Certain species of fish were abundant in some areas and scarce in others, so a network of trade and food exchange developed – kaihaukai. People would travel huge distances in large convoys from the coast, carrying dried seaweed, sharks and crayfish to trade for freshwater eels, kōura and mussels.

However, the unequal distribution of sea creatures across the landscape could also lead to bitter rivalries, and fights could break out over rich fishing grounds. Sometimes these disputes turned ugly, and the bones of an enemy might be carved into a fishing hook – an ultimate act of desecration that needed to avenged. To prevent fighting breaking out, ownership of areas of the ocean, rivers and lakes was carefully marked out, often with carved posts. Everyone was taught to respect these boundaries and fish only in areas where they had a right to. When fishing at sea, fishermen knew how to triangulate their position from landmarks on the mainland, and could identify hundreds of fishing grounds up to 40 kilometres offshore.

A scene from Taupō pā near Porirua. Artwork by George French Angas. *(Alexander Turnbull Library, PUBL-0014-57)*

Fish and shellfish were so vital to Māori life in Aotearoa that the act of fishing itself became sacred. The origins of fishing could be traced back to a battle between the atua – spiritual deities who were the embodiment of the natural world. Tāwhirimātea (atua of the wind and storms) waged a war against his brothers for separating their parents, the earth and the sky. Tāne Mahuta (atua of the forest and birds) escaped from the onslaught by hiding in the forests, while Tangaroa (atua of the oceans and sea creatures) hid at the bottom of the sea. Only Tūmatauenga (atua of war and the ancestor of humans) was able to withstand the winds and storms. For his bravery he earned the right for his descendants to harvest his brothers' offspring – to collect the fruit and birds of Tāne Mahuta and to harvest the fish and shellfish of Tangaroa.

In going fishing, Māori continued this long tradition. But because fish and shellfish were the children of Tangaroa it was essential that the correct protocols (tikanga) be followed and permission gained through karakia (prayer) before harvesting them. The entire act of fishing was considered tapu (sacred) and only those who were ritually pure themselves could partake in it. Eating would break the tapu, so no one on a fishing trip was allowed to eat until safely back on land. Fishing vessels, hooks and lines were all declared tapu, and when a large fishing net was being woven, the surrounding area and shoreline all became tapu as well. To ensure that Tangaroa would continue to look on the fishers favourably, the first fish would always be returned to the sea and, after the fishers arrived safely back on shore, seaweed and fish offal would be offered to Tangaroa, with thanks for a successful journey.

A waka sails past Mt Taranaki. Artwork by George French Angas. *(Alexander Turnbull Library, PUBL-0014-57)*

Fish and shellfish were so fundamental to Te Ao Māori that the sea came to be seen as the source of everything of value in Aotearoa. The art of whakairo (wood carving) was believed to have been discovered in the meeting house of Tangaroa at the bottom of the ocean. Pounamu (greenstone) and tūhua (obsidian) were thought to have once been fish that had swum to Aotearoa from the ancestral homeland Hawaiki.

Perhaps the most valuable thing to emerge from the sea was Aotearoa itself. The demi-god Māui was an expert fishermen, inventing a wide range of hooks, traps and fishing techniques. But his greatest catch of all was the North Island of New Zealand – Te Ika a Māui – which he was said to have caught with a magical hook formed from his grandmother's jawbone. The hook lodged in the fish at Hawke's Bay, and when he hauled it out of the water its full form could be seen: its fins flaring out to Taranaki and East Cape, the head of the fish at Wellington, and the tail stretching to Cape Reinga.

For Māori, fish and shellfish were inseparable from life itself. They set the rhythms of daily life, and gave it flavour and richness. Fish and shellfish were a source of identity, something that showed you were connected to a place and could provide for yourself and others. They were a valuable trading commodity as well as a cause of war and bloodshed. They were a link with the spiritual realm, and something to be treated with respect, wonder and awe.

NEW ARRIVALS

Across the waves, a new set of explorers came to encounter New Zealand's sea life for the first time. The first Europeans to sail through New Zealand waters were Dutch merchants, led by Captain Abel Tasman, in 1642, but their visit was brief and they didn't record any observations of natural history. Over a century later, however, the crew of HMS *Endeavour*, led by Lieutenant James Cook, were part of the greatest planned scientific voyage that the world had yet seen. Their goal was to observe the transit of Venus and record observations of the people, plants and animals they encountered on their travels in the Pacific.

Leading the study of natural history were Joseph Banks and Daniel Solander, students of the Age of Enlightenment who were driven by a passionate desire to uncover the mysteries of nature. Under their influence, the *Endeavour* became a floating laboratory, equipped with state-of-the-art scientific technology and an entourage of assistants, painters and illustrators. To study New Zealand's sea life, they came armed with an arsenal of nets, dredges, hooks, spears, glass bottles of all shapes and sizes, casks of spirits, and even a telescope for looking underwater.

Arriving in late 1769, the crew were overwhelmed by the explosion of life in New Zealand waters. Banks noted in his journal: 'Every creek & corner produces abundance of fish, not only wholesome, but at least as well tasted as our Fish in Europe … as many were caught with hook & line as the People could eat'.[1] Often before they had even dropped anchor they had already caught more than enough fish to feed the entire ship. As the *Endeavour* travelled around New Zealand the crew ate fish cooked in every possible way: boiled, stewed, grilled and baked into pies. They collected mussels from the rocks and dredged up oysters from the deep. The naturalists studied these sea creatures, painted them and preserved them in spirits. But sometimes the urge to see how they tasted became too great, and a number of specimens were half-eaten by the time they were preserved.

A scientific illustration of a crayfish (*Jasus edwardsii*) by John James Wild, 1885. (*Smithsonian Libraries*)

But perhaps what struck these explorers even more than the sea creatures themselves was the unique connection Māori had with them. As they travelled around New Zealand the crew described a land obsessed with kaimoana. Everywhere they went, people were collecting shellfish, cleaning and drying fish and diving for crayfish, and almost every house they visited had nets in the process of being woven. Māori seemed to possess some special knowledge about the sea, and it was always more profitable to trade with Māori than to try to catch fish. Cook admitted in his journal, 'We were by no means such expert fishers, nor were any of our methods of fishing equal to theirs'.[2] For their part, Māori seemed rather unimpressed with the newcomers' fishing skills; Banks recalled how in the Bay of Islands a group of Māori openly laughed at the small European fishing nets.[3] Rubbing salt into the wound, the group displayed their own enormous net, which was 10 metres deep and stretched almost a kilometre in length.

Sea creatures were often a source of connection between these two peoples from opposite sides of the planet. During early meetings the threat of violence was ever present, but when Māori offered up fish to the Europeans, usually the hostile atmosphere disappeared and a flurry of trading back and forth ensued.[4] Te Horetā Te Taniwha was a child living at Whitianga when the *Endeavour* arrived, and he believed the crew were tupua – supernatural creatures similar to goblins that were tapu and unable to eat human food. But when he saw the crew eating oysters and cockles he realised they must be humans after all.

Ship Cove in Queen Charlotte Sound, painted by John Webber during Cook's third voyage to New Zealand in 1777. *(Alexander Turnbull Library, B-098-015)*

THE NEXT WAVE

Fish and shellfish would become critical to the survival of the first Europeans who attempted to live in New Zealand. Sealers and whalers, stranded on the coast in remote areas for months or years at a time, had to learn how to make the most of the resources of the sea. They would catch hāpuku from the shore and collect oysters and mussels from the rocks, pickling them in casks for the cold winter months. The settlers of new towns such as Auckland and Dunedin relied upon resources brought in by Māori, and fish such as snapper and barracouta were often key to keeping these early settlements fed.

European explorers such as Charles Heaphy and Thomas Brunner learned from their Māori guides how to rely on native fish and shellfish as they traversed the countryside. They learned how to harvest eels from the river, to cook kākahi with raupō shoots, and how to net upokororo and cook them with the sweet roots of the cabbage tree. These early adventurers were keen to give any food a go, dining on slimy hagfish and even taking a particular liking to sea anemone, which Heaphy described as 'the most extraordinary food that ever afforded nutriment to the human body, and must be eaten to be comprehended'.[5]

Trading for fish on the southern shores of Hokianga Harbour. Artwork by Augustus Earle. *(Alexander Turnbull Library, PUBL-0015-04)*

But as more and more settlers arrived on New Zealand shores, many people turned away from the sea. The new arrivals saw themselves as building an agricultural paradise where every citizen could live like an English country lord.

Only wealthy aristocrats in Britain could eat beef and lamb for every meal, so this diet became the ultimate aspiration for many settlers. Fish, on the other hand, were associated with poverty, and native fish and shellfish came to be regarded with contempt. Pāua were seen as repulsive, snapper and kingfish were deemed coarse and tasteless, and crayfish were regarded as food for drunkards and the poor. For many settlers the only fish worth eating were British fish, and hundreds of tonnes of canned seafood were regularly imported into New Zealand.

So little value was placed on freshwater fish and shellfish in particular that many colonists believed New Zealand rivers were 'entirely devoid of inhabitants'.[6] In 1857 the politician Charles Hursthouse was particularly blunt, writing in a pamphlet encouraging British settlers to immigrate to New Zealand, 'Just as the New Zealand forests are destitute of game, so are its rivers destitute of fish.'[7] In a major national effort, with overwhelming public support, British fish were brought in to stock New Zealand's 'empty' rivers. Before the advent of refrigeration, fish fry were kept on ice and sailed across the globe to New Zealand, and all around the country rangers for acclimatisation societies travelled by foot and horseback to release trout into every river they could find.

Things took a darker turn for our native sea creatures as people realised there might be money in the sea. A fever to exploit New Zealand's fish and shellfish began to spread. The politician John Munro wrote in 1870 that 'there is an inexhaustible source of national wealth swarming unmolested round these islands'[8], while James MacAndrew addressed Parliament, telling the members that 'our waters are replete with golden sovereigns, and we have only to take them out and they will fill the Treasury'.[9] Fish were caught in such abundance that they frequently spoiled before they could be sold, with the excess buried, thrown into the sea or used as fertiliser for gardens. Unscrupulous fishers employed dynamite to target fish schools or poured sacks of lime into rivers and collected the dead fish along the banks.

In the twentieth century, huge technological advances in fishing methods helped to strip the ocean on an industrialised scale. Immense dredges trawled the sea floor, taking unbelievable hauls of seafood, but leaving destruction that could take centuries to heal. Some fish stocks never recovered and others, like the upokororo, disappeared forever.

In the face of such devastation, New Zealand began to lose not just a resource, but a spiritual connection with the sea that had thrived for centuries. As European settlers began to outnumber Māori for the first time in the 1860s, the roles around fishing began to change. For years Māori had dominated the fishing industry, with hundreds of waka supplying the ports of Auckland with fresh fish and oysters. But as more Pākehā participated in the fishing industry, the government created laws that effectively shut Māori out. Fresh waters that had fed people for centuries were drained to turn into farmland, and Māori were banned from catching native fish species, so that introduced trout had something to eat.

A TURNING OF THE TIDES

But even during this most destructive phase of our relationship with the sea, there were glimmers of hope. The invention of the Aqua-Lung spurred a new wave of underwater explorers to venture into the ocean depths. It was a wild new frontier, and early scuba divers applied a DIY mentality to their craft. They made their own regulators with old fire-extinguisher cylinders, and kept themselves warm by wearing jerseys, long-johns and sheets of rubber. These pioneers discovered new lands underneath the waves – incredible oases, like the Poor Knights Islands, bursting with ocean life. Until this point, learning about the sea had mostly relied on catching fish and shellfish and studying them out of their environment. But now these creatures could be viewed in their natural habitat, and knowledge of our oceans radically increased. Underwater photos and films helped to reveal the secret lives of our fish and shellfish to the public for the first time.

With a greater appreciation of our aquatic species, a new conservation ethos began to emerge. Tighter rules were brought in to manage fishing resources, and hatcheries once used to introduce exotic fish were turned into research centres to study native species. New Zealand became the first nation in the world to set up a no-take marine reserve, at Goat Island, north of Auckland, and, within a decade, reefs that had been stripped bare by kina were covered in lush forests of kelp. Huge schools of fish returned too, and hundreds of thousands of New Zealanders flock there every year to see them.[10]

New Zealanders even spread this conservation ethos to the rest of the Pacific Ocean, and took a bold stand against nuclear testing. When French secret service agents bombed the protest ship *Rainbow Warrior* in Auckland Harbour in retaliation, the wrecked ship was towed to Matauri Bay in Northland and turned into an artificial reef to promote sea life.

Today, fish and shellfish have become central to a shared New Zealand identity. Fishing and collecting shellfish are seen as rites of passage, and our seafood is now treasured and served in top-end restaurants all around the world. Fish and shellfish are depicted in innumerable ornaments, paintings and artworks, and have been written about in countless books, journals and magazines.

While there is still a long journey ahead, in many places Māori are regaining the right to protect and manage the ocean as their ancestors once did. Hapū have been working with scientists to use traditional Māori fishing techniques to study sea creatures, and Te Ao Māori has started to influence how many people relate to the sea. There is a growing awareness that the oceans provide people with more than just food and that when fish and shellfish thrive, people thrive as well.

The problems facing our oceans can seem overwhelming. Our native fish and shellfish face many threats, from climate change to plastic pollution, invasive species, habitat destruction and overfishing. But throughout New Zealand's history, the more people have learned about the sea, the more they have fought to protect it. By telling the stories of the sea, we reveal that our lives on land are intimately connected with the creatures that live around our shores.

There's a whole world out there beneath the waves, full of incredible species and remarkable stories. And there are still many more secrets waiting to be discovered.

John Dory (*Zeus faber*) by Edward Donovan, c.1808. (*Biodiversity Heritage Library*)

FRESH WATER

Creatures of the rivers, streams and lakes

Eels / Tuna

Slimy river monsters

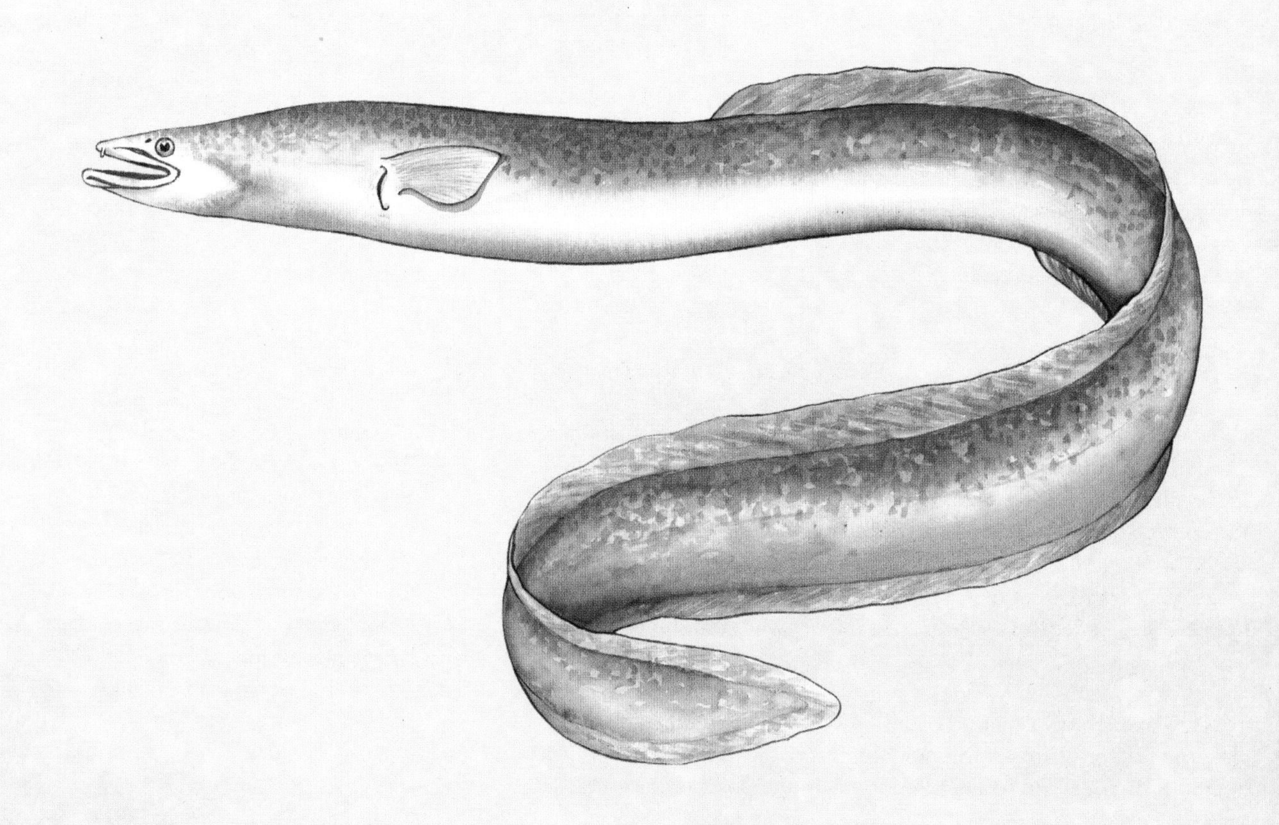

TAXONOMY

Three species of freshwater eel are found in New Zealand. The longfin eel (*Anguilla dieffenbachii*) lives only in New Zealand, the shortfin eel (*A. australis*) is native across the Pacific, and the spotted eel (*A. reinhardtii*) is a mostly tropical species that occasionally visits northern New Zealand. The longfin can be distinguished from the other species, as its dorsal (top) fin is much longer than its anal (bottom) fin, and when it bends there are large, baggy wrinkles in its skin.

New Zealanders have a complicated relationship with eels. At times they have admired them as the most valuable fish in the country, and at others they have viciously hunted them to near extinction. For Māori, eels represented one of the single most important food sources in Aotearoa, gigantic tubes of meat that could be found in almost every river. When thousands of eels made their annual migration (heke) downstream in March and April, it was perhaps the most abundant food source in the country.

Māori knew several ways of catching them: they could be speared with a sharp stick, shepherded into shallow water channels and caught by hand, or fished for with a piece of harakeke, using earthworms, grubs or glow worms as bait. But when the season was right, no other method could surpass the eel weir or pā tuna, which could catch thousands of eels.

These wooden fortifications were constructed against the flow of the river, and directed the migrating eels into a hīnaki (a woven trap). When the pā tuna were erected on the river, it was often a huge social event and the cause for much celebration. The heke only lasted so long, and it was important to make the most of this unrivalled abundance of food.

Anything that wasn't eaten right away was preserved for later. Harvested eels were smoked, sealed in kelp bags encased in boiling fat, or split in half and dried in the sun. They could even be kept alive in little pens in the river and fed bits of kūmara to fatten them up. When those hapū who lived near eel-fishing grounds hosted a feast, huge numbers of eels were expected by the visitors, and they were rarely disappointed. At a great feast held by Ngāti Hauā in Matamata in 1838, more than twenty thousand dried eels were served.

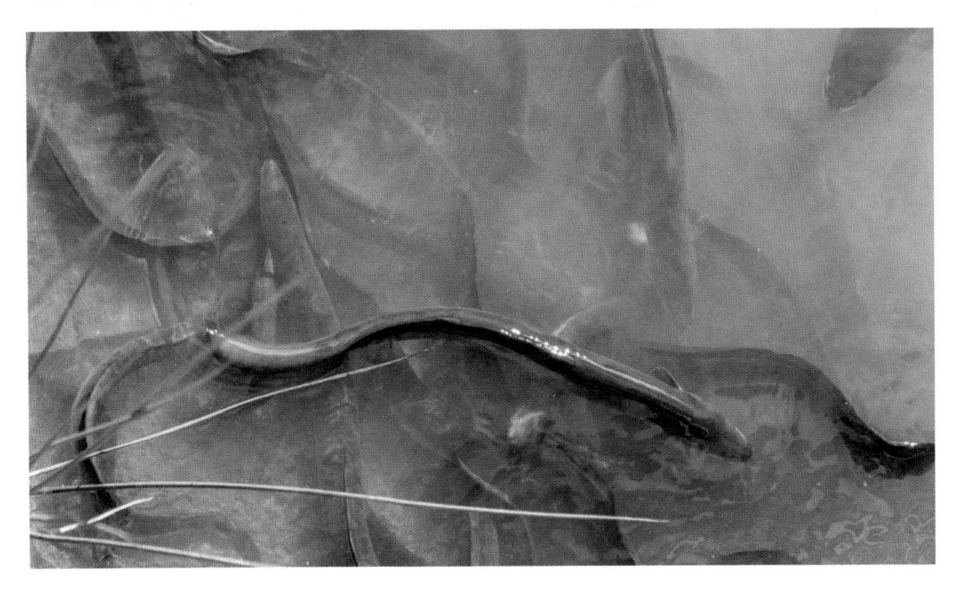

A swarming mass of eels migrating down a river. *(iStock)*

Spotted eel (*Anguilla reinhardtii*) by Frank Olsen. *(Department of Harbours and Marine, 1965; courtesy of Queensland Government)*

THE HEAD OF TUNAROA

Eels were considered such a significant food source that they were often seen as distinct from other fish and regarded as a separate category of food. Large eels were kept as pets or regarded as taniwha – powerful deities that protected the river and made sure everyone followed the proper fishing protocols.

Eels were thought to have a divine origin, emerging from the head of the eel god Tunaroa. When Tunaroa attacked Māui's wife Hina, Māui sought revenge. He dug a trench and lured Tunaroa into the shallow waters, before hacking him to pieces with an adze. The tail of Tunaroa was thrown in the ocean and became the sea-going conger eels, his nostril hairs became the supplejack vines of the forest, and his head fell in the river and gave rise to the freshwater eels.

A New Zealand longfin eel (*Anguilla dieffenbachii*) emerging from the shadows. *(iStock)*

A FOOD FOR TRAVELLERS

Pākehā soon recognised the value of eels, and learned from Māori the best ways to catch them. They were especially useful for early European travellers as they ventured into remote parts of the country. Nathaniel Chalmers, one of the first Europeans to explore Central Otago, was amazed that his Māori guides could supply the whole travelling party with as many eels as they could eat, using nothing more than wooden spears and hooked sticks. Another early adventurer, Thomas Brunner, would eat eels for three meals a day; whenever Brunner and his Māori guides ran low on food, they would simply stop at a nearby river to collect more.

Nineteenth-century German surveyor Gerhard Mueller relied on eels for food during his trips through South Westland and wrote, 'Prejudiced people have … shuddered at the very thought of eating such water serpents, but in a week or two they went back full of Eel and gratitude, confessed they had discovered a new pleasure.'[1]

An eeling party with their catch, c.1910. *(Alexander Turnbull Library, 1/2-028987-G)*

EEL DESTRUCTION

At the end of the nineteenth century, however, the Pākehā relationship with eels completely changed as they came into contact with the most beloved of all European fish, trout. An incredible national effort had gone into stocking New Zealand rivers with trout, but it was soon discovered that baby trout were being eaten by eels. Many people were suddenly overcome by a feverish desire to exterminate these predators.

With the support of government agencies, the nation's acclimatisation societies declared that eels were 'public vermin' and launched eel destruction campaigns in the 1930s to remove them for good. The societies hosted public eel dissections, displaying to horrified audiences eel stomachs full of trout. Farmers and fishermen around the country were encouraged to kill as many eels as possible and were given eel traps and paid cash bounties for every eel killed. Acclimatisation societies published accounts of the most effective methods for killing eels. Writing in 1933, Auckland Acclimatisation Society ranger James Dobson believed the best method was 'wading streams at night with a bright lamp and sabre … Where infestation is bad, it is possible to wade up a stream beheading the eels in one stride.'[2]

This fever for eel destruction eventually faded, once it was realised that eels actually benefitted trout populations. In areas where eels had been wiped out, the rivers quickly became overpopulated with small, undernourished trout that were no good for fishing. But just as eel destruction began to wrap up in the 1960s, a huge overseas market for eel meat was established, and the commercial eel fishery exploded. For a time, eel was the second-most valuable fish export in the country, behind crayfish.

ETYMOLOGY

The name *Anguilla* is the Latin word for eel, and the species name *dieffenbachii* honours the geologist-naturalist Ernst Dieffenbach, who sent specimens of New Zealand eels to the British Museum in the 1840s. The Māori name, tuna, is an ancient word with roots across Polynesia and Melanesia, where it refers to eels and eel-like fish. It's a sign of the importance of eels in Māori culture that there are over 100 different names for these creatures around the country, recognising different species, sizes, colours and life stages.

With so much money to be made, there were few limits or controls placed on the harvest. Eels of all size classes were taken, especially the enormous, slow-moving adults. However, these large eels could be over a century old and spawned only once, at the very end of their lives. The eel fishery is now under stricter management, but unfortunately, these repeated sustained attacks on the eel population and the removal and destruction of their habitat have taken their toll, and our endemic longfin eel is now in serious decline.

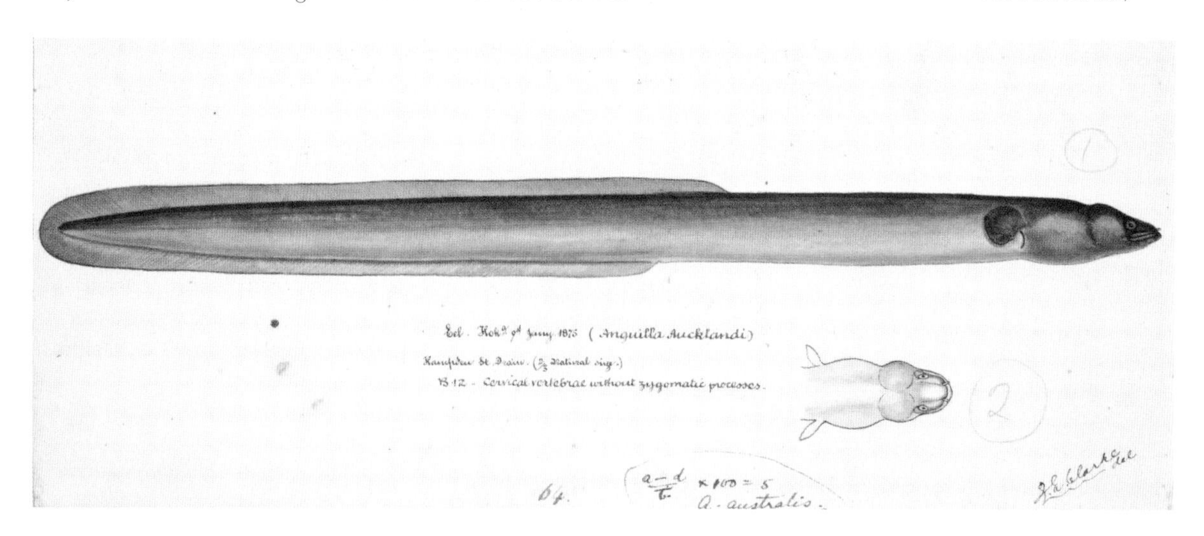

Illustration of a longfin eel by Frank Edward Clarke, 1875. *(Te Papa, 1992-0035-2278/17)*

A NEW APPRECIATION

Today the mood has once again shifted towards eels, and they have gone from being seen as a public menace to taking up a greater role in our national psyche. Many people who grew up in rural areas have fond memories of having a 'pet' eel in a river that they fed their scraps to. Some of these eels have even become local icons, such as Doris, a large longfin eel in the Avon River in Christchurch. Doris was so beloved that residents successfully campaigned for Christchurch City Council to establish an eel-fishing ban to protect her. New Zealand longfin eels have attracted international attention for their immense size, and have been featured on TV shows such as *Monster Fish* and *River Monsters*. They remain an important taonga species for Māori, sculptures and artworks of eels have become common around the country, and encounters with eels have become tourist attractions, allowing people to feed giant eels with scraps of meat.

BIOLOGY

New Zealand longfin eels can grow to be the largest eels in the world, weighing more than a small child, and can live for up to 100 years. At the end of their lives, they leave the fresh water and travel along the sea floor to a location in the Pacific Basin near Tonga, where they spawn. Their offspring – known as leptocephali – look like delicate, transparent ribbons, which are nearly invisible. For several months they swim back towards New Zealand. As they get closer to land they transform into transparent tubes known as glass eels. The glass eels swim upstream to find their new homes and begin to take on their adult forms.

Despite their reputation, eels can be quite alluring and charismatic creatures. *(SeacologyNZ)*

Lamprey / Kanakana

Vampires of the sea

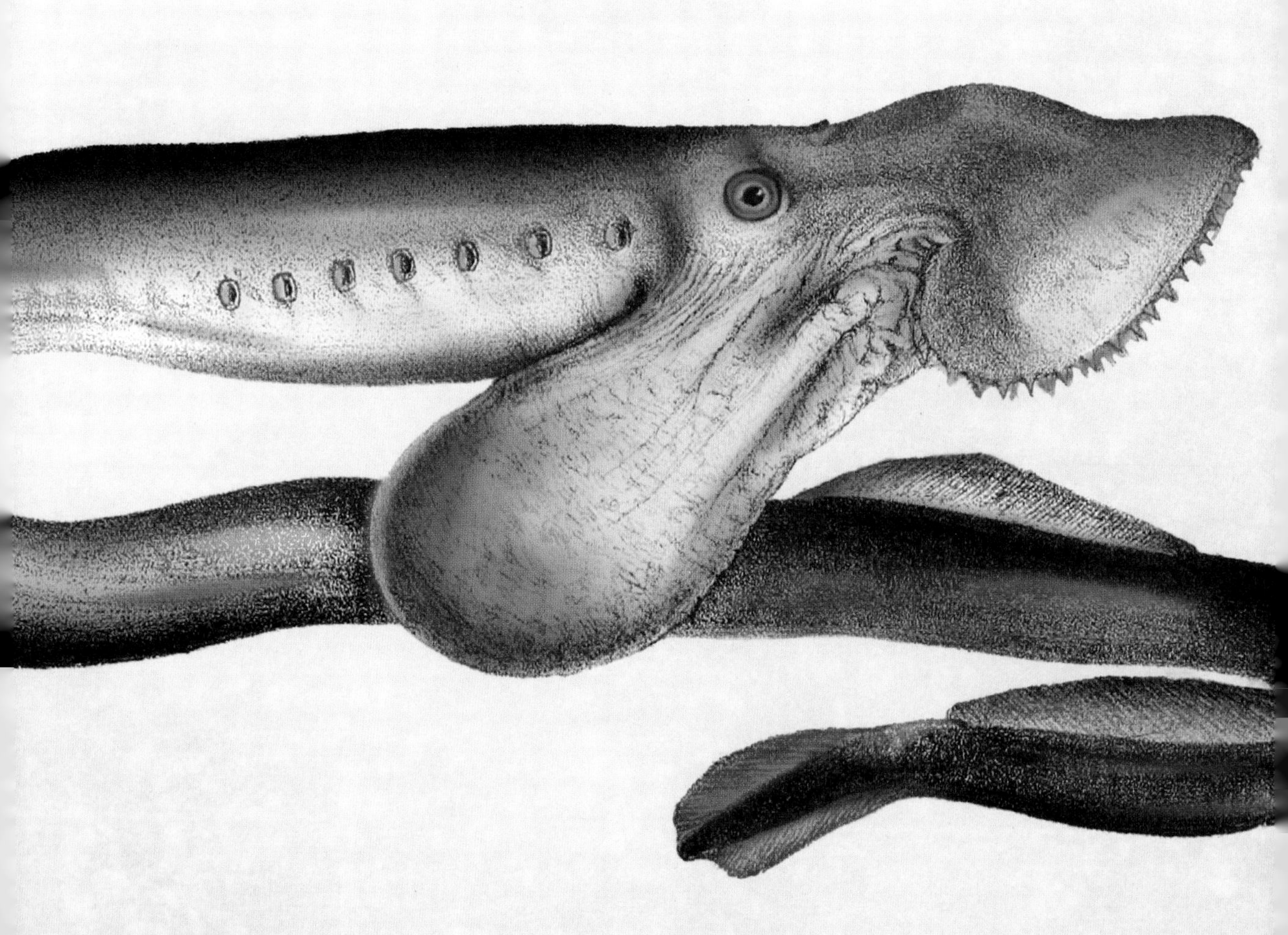

TAXONOMY

New Zealand has one species of lamprey, known as the pouched lamprey or kanakana (*Geotria australis*). Its sea-going and freshwater forms are so different that they were once thought to be different species. Lamprey are considered 'living fossils' and belong to an ancient order that diverged from all other backboned animals around 500 million years ago, long before the advent of the dinosaurs.

Lamprey are among the most bizarre fish on the planet. In fact, they are on the very border of what we might consider to be a fish. They have no jaws, no scales and no swim bladder. When they first hatch, they resemble blind worms, and lie buried in the sandy bottoms of freshwater streams, filter-feeding on particles that float by. But once they reach maturity, they undergo a radical transformation, growing a circular mouth with spiralling rows of teeth and leaving for the ocean, where they turn a striking electric-blue colour. Here they become parasites, using their powerful suction-cup mouths to latch on to fish and whales and sustaining themselves by drinking their hosts' blood. After several years of feeding, they detach and return to fresh water to spawn.

Their return journey is equally incredible. They travel upstream for months on end without food, sometimes traversing hundreds of kilometres. If they encounter an obstacle, they use their sucker mouths to climb over it, and can scale vertical waterfalls using only their mouths. When they are reproductively mature, females swell and become fat, while males grow a strange baggy pouch from their throat. When they finally arrive at the upper reaches of streams, they pair up underneath boulders to spawn and lay their eggs to begin the cycle again.

The sea-going form of lamprey by Frank Edward Clarke, 1876. *(Te Papa, 1993-0035-2278/22)*

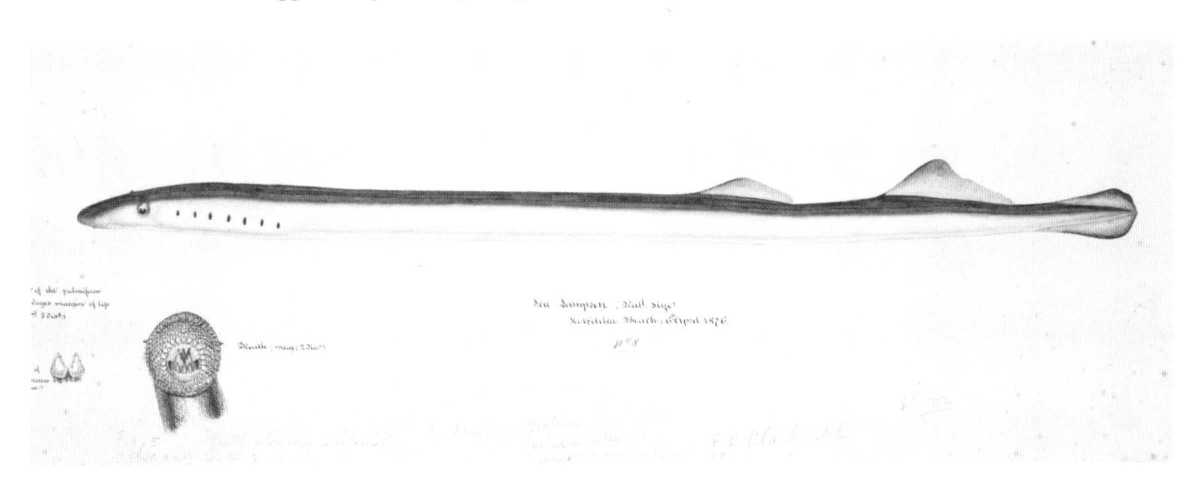

A SYMBOL OF PERSISTENCE

Māori admired the way lamprey didn't let any obstacle stand in their path, so someone with great endurance and stamina was said to have the heart of a lamprey: 'he manawa piharau'. The lampreys' annual appearance during Matariki made them an important food supply at the coldest and least productive time of year, when there were few other fresh protein sources available. During this time, groups of fishers might camp out at lamprey grounds for several months.

Pouched lamprey (*Geotria australis*); original artwork by W. Wing (1851), adapted by Lars Quickfall.

There were a number of methods for catching them. One of the easiest was to collect the fish as they were ascending a waterfall. All that was needed was to place a bag underneath them and disturb them from above, or lean out and pluck them off the rock face. In wider rivers, a different tactic was adopted – the utu piharau. These wooden structures were designed to block the lamprey from moving upstream. As the fish tried to make their way through gaps in the structure, they were washed back by the force of water into carefully placed hīnaki (woven traps). Utu piharau were once found all along the Whanganui River, with each pā and settlement having one. Some could be over 15 metres long, with multiple hīnaki. Often so many lamprey were caught that they couldn't be carried away from the river and had to be transported on purpose-built rafts.

ETYMOLOGY

The origin of the name *Geotria* is unclear but it may come from the Greek *Geotragia*, which means 'eating of earth', a possible reference to the lamprey's suction mouth. The name *australis* means 'southern'. One suggestion for the origin of the word 'lamprey' is that it derives from Latin words meaning 'stone lickers', a reference to the way some lamprey species use their suction mouth to build nests with stones.

'Kanakana' is the most common Māori name used for lamprey in the South Island, and is said to mean 'darting about', reflecting the fish's erratic swimming style, while in the north it is known as 'piharau', a reference to its many (rau) gill slots (piha).

Iwi that were lucky enough to inherit rights to lamprey-fishing grounds could expect to gather several tonnes over a season, and feasts where thousands of lamprey were served were not uncommon. Their rich, oily flesh was considered a delicacy across

An utu piharau on the banks of the Whanganui River, sometime in the late 1800s. *(Alexander Turnbull Library, 1/1-000482-G)*

the country, and they were traded widely with iwi on the coast for mutton-birds and seaweed. Lamprey were eaten roasted, grilled, steamed or boiled, and were often served with bracken fern as a relish. It was best to catch them as soon as possible after they entered fresh water and before the pouch formed on the males, as that is when they are fattest and richest in flavour.

A SURFEIT OF LAMPREY

Early European settlers described incredible numbers of lamprey in the rivers, with so many climbing up waterfalls that you could simply take off your hat and scoop them into it. In 1858, Southland explorer Nathaniel Chalmers witnessed a giant column of lamprey over a kilometre long travelling up the Mataura River that was so tightly packed that at first it was believed to be a single giant eel.[3] And in the 1930s, so many lamprey would sucker onto the water wheel at the Mataura Falls they could regularly bring it to a standstill.[4]

While British settlers could be quite squeamish about eating New Zealand seafood species, the strange-looking lamprey were generally well liked by those who tried them. Europeans admired their fine flavour and tender flesh with very few bones, describing the taste as resembling sardines or salmon. A number of settlers eagerly traded with Māori to secure baskets of lamprey.

Another part of the appeal was that lamprey had a long history as food in England. Potted lamprey were considered a delicacy and associated with the nobility, and English schoolchildren grew up learning about King Henry I, who was said to have died from eating too many lamprey in a pie. While this story sounds apocryphal, there were reports that eating too many lampreys was a cause of death in New Zealand as well. Reverend Richard Taylor, writing in the 1850s, reported that deaths from feasting on lamprey were 'by no means uncommon'[5], and other sources note that rangatira were most at risk, as they were given the largest portions as a sign of their status.

Elsewhere there is evidence that Māori recognised the dangers of eating lamprey, and that care was taken when cooking them in hāngī as a dark fluid could be expressed from their skin that could make people ill. There might be something behind this claim, as lamprey are very 'primitive' vertebrates and lack many of the specialised organs and functions of higher animals. Because of this they are unable to excrete some of the toxic compounds that accumulate in their skin.[6] If enough lamprey are eaten, then sufficient toxic compounds might be consumed to receive a lethal dose.

BIOLOGY

Very little was known about lamprey breeding until recently, when New Zealand scientists observed male and female lamprey pairing up and making nests underneath boulders. After spawning, the pairs spent several weeks guarding and caring for their eggs. The male uses the large baggy pouch on its throat to gently brush the eggs and keep them oxygenated. It is thought that adults are guided to spawning sites by detecting pheromones in the water. Sadly, lamprey population sizes have decreased over time, and the species is now considered threatened in New Zealand.

Kākahi / Freshwater mussels

Parasitic hitch-hikers

A live kākahi on the bottom of a lake, showing its extended siphons. (*© EOS Ecology / www.eosecology.co.nz*)

TAXONOMY

There are three species of kākahi, all endemic to New Zealand. The most common is *Echyridella menziesii*, which is found across New Zealand. *E. onekaka* are found only in the northwestern South Island, and *E. aucklandica* has a strange distribution in the lower North Island and Southland, where it is thought to have been transported by Māori. Kākahi belong to the freshwater mussel family, Hyriidae, which is found across the southern hemisphere, and have likely been present in Aotearoa since its separation from Gondwana, over 80 million years ago.

Kākahi are muddy-brown mussels that spend their lives hiding out on the bottoms of rivers and lakes. They are fairly easy to overlook, and yet these unassuming creatures were once one of our most celebrated shellfish.

Kākahi were eaten from the very earliest days of settlement in Aotearoa, and the shells are a common feature of Māori middens, sometimes found intermingled with moa bones. Prolific kākahi beds can produce several kilograms of mussels in a single square metre, and were probably one of the reasons some iwi decided to settle around the central North Island lakes. Kākahi were a reliable food source that could be collected year-round, and if guests arrived unexpectedly, it was easy to go out and dive for some mussels to provide food at a moment's notice.

To collect large quantities of mussels, the most effective tool was the rou kākahi or mussel dredge. Dredging was conducted with great ceremony, with the lead dredger standing at the stern of the waka adorned in a dog-skin cloak. As the waka moved about the lake, the dredger would move the rou kākahi handle with thrusts and parries, like a warrior in battle. In this way the waka could be filled with kākahi in little time at all. Mussel dredgers were seen as highly skilled and energetic, and it was believed they made good husband material. In a popular whakataukī (proverb), women were encouraged not to put up with a lazy man and to seek out a mussel dredger instead:

Tāne moe whare, kurua te takataka;
Tāne rou kākahi, aitia te ure.

A man drowsing in the house, smack his head;
A man skilled in dredging kākahi, marry him.

Kākahi (*Echyridella menziesii*) by Arthur Powell (1947), adapted by Lars Quickfall, alongside one of the species' host fish, the smelt (*Retropinna retropinna*), by Frank Edward Clarke (1890).

THE JUICE OF THE KĀKAHI

Kākahi could be eaten raw or lightly cooked in a boiling-hot spring, with care being taken to remove them from the heat before they became too rubbery. Dried kākahi were used as food when travelling, or added to stews and served with raupō roots. The meat was believed to have healing properties and was used as a natural remedy, after being softened with water and made into a soothing broth – wai kākahi – that was easy to digest. This broth had a particularly important role in child rearing and was fed to young children, especially those who did not have a mother to nurse them. When babies wouldn't stop crying, the mussel meat could be given to them to suck like a dummy.

Many whakataukī reference the important role kākahi played in child rearing. It was said that 'Ko te kākahi te whaea o te tamati' – 'The kākahi is the mother of the child' – and someone who had been raised with a proper understanding of Te Ao Māori was said to have been 'suckled on the juice of the kākahi'.

Since the early twentieth century kākahi have been used to make a curry or a chowder with milk. Immersing them in salt water overnight is said to improve their taste, as it makes the mussels spit out any mud and silt and take on a saltier flavour.

ETYMOLOGY

The name *menziesii* honours Archibald Menzies, the surgeon aboard HMS *Discovery*, under Captain George Vancouver, which visited New Zealand in the 1790s. The word 'kākahi' derives from the Polynesian word 'kahi', which is used for a number of bivalve shellfish in Polynesia. The mussels are known by a number of different names across the country, such as 'torewai' around Lake Ōmāpere and 'kaeo' elsewhere in the north. A number of places are named after this shellfish, including the town of Kaeo and Lake Rotokākahi (the Green Lake), near Rotorua.

AN ACQUIRED TASTE

Not all iwi valued kākahi, and in a number of places they developed a reputation for being bland and unappetising. One traditional account from the Central North Island says that when the rangatira Te Rangitaumaha sent baskets of kākahi as a present for the feast of his grandchild, the child's mother, Te Huhuti, was upset and considered it to be an insulting gift. In some versions of the tale, Te Huhuti ran away in shame, renouncing her claim to her ancestral lands, while others say that Te Rangitaumaha sent his children away to work for Te Huhuti and produce food for her to try and make amends.

European settlers were rather unimpressed with the taste as well. Ethnographer Elsdon Best described kākahi as 'very insipid', and early North Otago settler Sherwood Roberts wrote, '"Frightful" … best describes their flat and nauseating flavour.' However, some Pākehā came to rely on them in the same way that Māori had, as a quick, easily procured food source. Thomas Brunner, who travelled extensively with Māori along the west coast of the South Island in the 1840s, ate them regularly and believed they made a perfectly palatable dish, especially when served with the roots of raupō.

RATTLING SHELLS

As well as food and medicine, kākahi shells provided another useful resource. Their edges could be sharpened and used for splitting flax leaves for weaving and scraping kūmara and other vegetables. They could also be used to cut hair or to cut the umbilical cord after childbirth. There is even a tradition that the taniwha Hotupuku, which lived on the Kāingaroa Plains, was slain by a man named Pitaka, who used kākahi shells to slash open his stomach.

A bunch of kākahi shells linked together makes a rattling sound, and such rattles were used in toys and children's kites. They could also serve a practical function: in some places, strings of rattling kākahi shells were used to keep rats out of kūmara plantations – bunches of shells were tied to flax ropes, which were strung up over the plantation in every direction. Men stationed on night watch would periodically pull the ropes, causing the shells to rattle and frighten the rats away.

Kākahi have a polarising taste: some people regard them as delicious while others find them bland or nauseating. *(Alton Perrie)*

TINY PARASITIC HITCHHIKERS

It would be easy to assume that kākahi live fairly uninteresting lives – but that couldn't be further from the truth. In fact, for a part of their lives kākahi are parasites on fish. In their early larval stage, kākahi are known as glochidia, and resemble tiny clams with fangs. They use these 'teeth' to cling on to passing fish, hitching a ride for several weeks before dropping off upstream and developing into juvenile mussels.

Different species of kākahi live together in the same area, so to avoid competing with each other they have evolved to target different fish. The most common kākahi species (*Echyridella menziesii*) is not fussy, broadcasting its glochidia into the water column and sometimes wrapping them in a web of mucus to entangle passing fish, such as the common bully (*Gobiomorphus cotidianus*) or kōaro (*Galaxias brevipinnis*).

On the other hand, the highly endangered *Echyridella aucklandica*, is incredibly specific and has come up with an elaborate system to ensure only the right species of fish picks up its glochidia. To do this, it wraps them in special packages that somewhat resemble freshwater leeches or mayfly larvae, possibly as a ruse to trick fish into trying to eat them. So far, the only fish known to transport *E. aucklandica* is the smelt (*Retropinna retropinna*) and, even then, very few glochidia successfully attach. One possibility is that the primary host was originally the extinct upokororo and the loss of this fish could be partly responsible for the decline of this species of kākahi.

Adult kākahi move about with a muscular foot in a similar way to saltwater bivalves like toheroa and tuatua. *(Alton Perrie)*

HUMBLE HEROES

With kākahi in decline around the country, scientists are now beginning to understand the critical role they play in cleaning up freshwater ecosystems. Because adult kākahi feed by filtering the water column for bacteria, phytoplankton and bits of animal matter, they contribute to keeping freshwater environments clean. It's estimated that a single kākahi can filter a litre of water in an hour, meaning that a healthy population could filter the entire volume of a small lake in just a few days. This cleaning ability is so effective that some people have used kākahi in aquariums and ponds to remove algae.

Research is now focusing on growing kākahi in the lab and using traditional Māori translocation techniques to reintroduce them into waterways, in the hope that stable populations might help clean up our rivers and lakes.

BIOLOGY

Kākahi move about using their muscular foot, and burrow into the sand and gravel to anchor themselves. They have different feeding strategies as they grow: as larvae they are parasitic on fish, as juveniles they feed by grabbing food with their muscular foot, and as adults they filter bacteria and phytoplankton out of the water.

All three species of kākahi are threatened, and in order to protect them, the whole ecosystem needs to be protected. To thrive, kākahi require healthy populations of native fish; open passages for the fish to travel through; clean, clear water free of weeds for young kākahi to grow; and pest control on land to stop rats and other predators from eating the adult shells.

Kōura / Freshwater crayfish

Creepy crawlies

TAXONOMY

There are two species of endemic freshwater crayfish in New Zealand. The northern kōura (*Paranephrops planifrons*) is smaller than the southern kōura (*P. zealandicus*), which also has much hairier fighting claws. DNA studies suggest kōura may have been present in New Zealand when the landmass was still part of the supercontinent of Gondwana, over 80 million years ago.

Kōura were highly valued as a food source by Māori, and those who did not live near a lake or river would travel great distances to trade with other iwi to secure them. While these freshwater crayfish were small, they more than made up for their size in their abundance. Huge swarms were said to cover the bottom of lakes, as thick as bees in a hive. Iwi living in the central North Island were surrounded by bountiful kōura populations, and could hold great feasts with the crayfish as the main dish. At one gathering at Awahou, near Rotorua, in 1899, 600 guests stayed for a week, feasting almost solely on kōura.

Kōura could be eaten raw, roasted over a fire or boiled by dunking them in a geothermal hot pool. Another preparation method was to leave them in a river until they began to soften and could be slipped out of their shell in one piece then squashed flat and dried. These dry crayfish cakes were small, thin and light and made a useful snack when travelling.

TAU KŌURA

There were a number of ways to catch kōura using nets, pots and dredges, or by reaching into their underwater burrows and pulling them out by hand. Kōura diving – ruku kōura – was a popular pastime for both adults and children, and sometimes involved diving down more than 5 metres. But perhaps the most effective method was the tau kōura. Bunches of dried bracken fern were bundled together with vines and placed at the bottom of the lake or river. Kōura were drawn to these clumps of fern as somewhere to live, and to feed on the fern spores. After several weeks, the fern bundles were carefully drawn into a net and pulled out of the water. In this way, a waka could soon be filled with hundreds of kōura madly flapping about.

Kōura were an important food source and were once the mainstay of immense feasts. *(SeacologyNZ)*

Northern kōura (*Parenephrops planifrons*) by William Wing (c.1844), adapted by Lars Quickfall. *(Biodiversity Heritage Library)*

Given that kōura was such an important food source, it's unsurprising that good kōura grounds were highly coveted, and even good-quality ferns for kōura fishing were sometimes a source of jealousy and contention. To make sure that everyone followed the rules, wooden pou (posts) were used to mark out fishing grounds, carved so that everyone knew which family or hapū was allowed to fish there. A network of wooden pou once covered the entire lake of Rotorua, with some pou even having their own names and a long history associated with them. However, even with such regulations in place, poaching could be a real problem, and some fishermen chose not to mark their best spots so their kōura couldn't be found by others.

ETYMOLOGY

The name *Paranephrops* means 'like *Nephrops*', a species of marine lobster, and the name *planifrons* means 'flat forehead'. The word 'kōura' originated in Polynesia, where it is used for both freshwater crayfish and spiny lobsters, as it is in Aotearoa. Kōura are also known as kourawai, kēkēwai, karawai, koeke and kēwai, especially in the South Island, and are sometimes affectionately known as 'crawlies'.

It's believed that Māori may have actively managed kōura populations, moving them about various waterways to keep stocks up and introduce them into new habitats. There are traditions in the South Island that kōura were placed in ponds and lakes marked with cabbage trees, so that travellers making the journey across the Southern Alps to gather pounamu would always have access to this source of food.

CATCHING CRAWLIES

Many European settlers were fond of kōura, and while some thought the little crayfish tasted muddy, others found them delicious. Explorer John St John thought they held huge potential, writing in 1873, 'Delicious as the [kōura] is when boiled, or steamed in a hangi au naturel, or even when knocked-up in a curry by the untutored paws of a bush cook, what will he not be when science has stepped in, when he has been made the subject of skilful gastronomic experiments.'[7] Early settlers sometimes relied on kōura as an easy, quick meal, and soldier Gilbert Mair boiled them in hot-water springs and ate them for lunch while touring the Pink and White Terraces. In the early twentieth century, curried kōura tail was a popular item on dining-room menus, and they were served to the Prince of Wales when he travelled to Rotorua in 1920.

Catching 'crawlies' was a favourite sport enjoyed by children who grew up in the countryside, near forested streams. Kōura could be lured into the shallows using worms as bait, or the bravest children would stick their hands into a den while trying to avoid getting nipped. In Christchurch in the 1920s, kids made pocket money by selling captured crayfish to local jewellers, who used the gastric mill – a feeding apparatus that crayfish use to grind their food – to make jewellery.

Unfortunately, kōura are no longer as common, as their habitats have been severely altered, and their tough shell proves little match for introduced trout, which can swallow them whole. Trout that have been feeding on kōura have a pink, orangey flesh that was once highly valued, and kōura were sometimes introduced to provide a food source for trout, such as in 1926 when thousands were taken from Lake Rotorua and introduced to the Arapuni hydro-lake in the Waikato. Trout have such a fondness for kōura that some fly-fishing lures are designed to mimic the action of the little crayfish scuttling away.

KŌURA CUISINE

In recent years, there has a been a resurgence of interest in kōura. In Southland pine forests, empty lakes that have very little biodiversity value have been used as breeding pools to build up populations of southern kōura (*Paranephrops zealandicus*) to sell as food. These farmed kōura are served by chefs at a number of high-end restaurants, and have received a range of food awards. They are served boiled, barbecued or chargrilled with a blow torch, and their flesh is described as delicate, sweet and delicious.

There have been advances in our understanding of their biology as well. For years scientists struggled to gather a clear picture of kōura populations, but recently have been embracing traditional Māori fishing methods. Scientists working with iwi have been using tau kōura to scientifically survey kōura populations, finding it far cheaper and more effective than any other method of gathering the crayfish.

BIOLOGY

Kōura mostly move about at night, scavenging plants and invertebrates from the bottoms of lakes and rivers. They have been known to live in swamps that dry out over summer, by burrowing down into the mud and waiting until the water returns.

Unlike native saltwater crayfish, kōura have large pincers, which they use to both fight off predators and catch food. Despite their tough shell they are eaten by a number of native and introduced species such as eels, giant kōkopu, shags, trout, perch, catfish and even rats.

Unlike saltwater crayfish, freshwater kōura have large claws that can give a painful nip. *(Shaun Lee)*

Whitebait / Īnanga

A slippery delicacy

Banded kōkopu (*Galaxias fasciatus*), kōaro (*G. brevipinnis*), giant kōkopu (*G. grandis*) and īnanga (*G. maculatus*) by Joseph Drayton, Frank Edward Clarke and James Francis McCardell. *(Te Papa)*

With most fish, it is the adults that are most recognisable, but in the case of whitebait it is the small translucent juveniles that have become a national icon. For Māori, whitebait were an invaluable food source, and on the South Island's West Coast they could be caught in such abundance during the season that it was possible to live on whitebait alone. To catch these slippery delicacies as they migrated from salt water to fresh water in the spring, fishers would stand on the shoreline and scoop them up with nets, or establish permanent nets across a river mouth. The fish were either dried in the sun, roasted over a fire or steamed in a hāngī, and sometimes mashed together to form a patty. Once cooked, they were served with fresh fern fronds and pūhā, and garnished with the fragrant leaves of the heketara – tree daisy.

The translucent juveniles of the various whitebait species are very difficult to tell apart. *(iStock)*

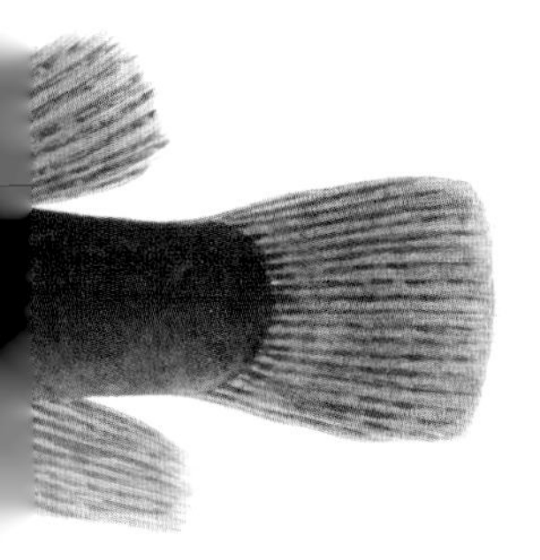

Those whitebait that escaped the nets would grow into five remarkably different adult fish: banded kōkopu, shortjaw kōkopu, giant kōkopu, kōaro and īnanga.[8] The adult fish were highly admired for their sweet taste and rich, fat flesh, and unlike the migrating juveniles they could be harvested year-round. They were especially useful as food for travellers, as they could be found in almost any forested stream around the country. Most species were hunted by torchlight at night, as this is when they swim to the surface to search for insects that have fallen onto the water.

In some places, like Te Urewera, women were regarded as better than men at catching kōkopu. They would walk upstream with a fish basket on their hips and a net in hand, churning up the muddy river bottom behind them to encourage the fish to investigate. Then they would spin around, deftly manoeuvring the fish into the net with their feet as they walked back down the river. The fish could also be caught by tying a huhu grub to a piece of flax and dangling it in the water as bait, by sweeping them up in nets, by spearing them with a sharp stick, or even scooping them out of the water by hand.

The different species of fish were recognised by Māori, with different traditions and beliefs surrounding them. For instance, the banded kōkopu was especially admired for its beauty. The name kōkopu was used for beautiful forms of pounamu, and someone exceptionally beautiful might be compared to the fish by saying they were *Me he kōrinorino kōkopu* – like a mottled kōkopu. The kōaro, on the other hand, was very important to the people of the central North Island, as the fish are exceptional climbers and can travel hundreds of kilometres inland up rivers. Kōaro formed a key part of the diet in these areas for several months of the year, and their sweet, nutritious flesh was said to help mothers produce milk.

TAXONOMY

The group of fish collectively known as 'whitebait' include five members of the Galaxiidae family that all migrate to the sea after hatching and return to fresh water to spawn. The migrating galaxiids are īnanga (*Galaxias maculatus*), banded kōkopu (*G. fasciatus*), giant kōkopu (*G. argenteus*), kōaro (*G. brevipinnis*) and the shortjaw kōkopu (*G. postvectis*). There are also twelve galaxiid species that don't migrate to sea and never leave the river or lake they were born in. Many of these species have probably been present in New Zealand since before the breakup of the Gondwanan supercontinent.

THE NATIVE TROUT

There were some European settlers who enjoyed these little fish, and made a hobby out of catching them with fly-fishing rods and lures that mimicked the fish's insect prey. Biologist David Graham fondly remembered catching kōkopu as a boy in the 1920s, and regarded it as the best fish he had ever tasted. But, on the whole, European settlers regarded kōkopu or 'native trout' as a poor substitute for the brown trout of their homeland.

The mirror image of a banded kōkopu feeding on insects that have fallen onto the surface of the water. *(SeacologyNZ)*

When brown and rainbow trout were introduced to New Zealand from the 1860s onwards, they became a significant predator of adult and juvenile native fish, and competed with them for food. Introduced trout were given extensive legal protection and their stocks monitored and topped up as needed, while native kōkopu were regarded as little more than trout food. In some areas Māori were banned from fishing for kōkopu to ensure trout had plenty to eat.

WHITE GOLD

The juvenile fish, on the other hand, were a completely different story, and 'whitebait' would become one of New Zealand's most valuable catches. Early adventurers described the incredible abundance of these fish in the early days of settlement, with Thomas Brunner writing that on the West Coast in 1850 'they are in such shoals that I have seen the dogs standing on the banks and lapping them from the stream'.[9] During the gold rush of the 1860s, diggers relied on catching whitebait to supply them with food on the remote West Coast, and Chinese gold-miners even made money on the side by drying whitebait and exporting it to China. In these early days, supply often exceeded demand, and cartloads of fish went to waste before they could be sold. Excess whitebait was fed to the chooks or used as cheap fertiliser for gardens. In 1899, ichthyologist Frank Clarke reported seeing acres of gardens covered with several inches of whitebait.

In the twentieth century, the price of whitebait began to rise, and whitebait came to be seen as a luxury item reserved for special occasions. One popular song of the 1930s even had the refrain 'When we get married, we'll have whitebait for tea.' With the price of whitebait skyrocketing, Pākehā fishers flocked to remote West Coast rivers in search of the fish that came to be known as 'white gold'.

In these isolated communities, resourcefulness was essential: nets were improvised from supplejack vines and mosquito netting; cabbage-tree trunks were turned into spotter boards, which were placed on the river bed so fish could be seen swimming into the net; and fish were stored in old kerosene tins. Other entrepreneurs hacked airstrips out of the Fiordland bush so they could collect whitebait by plane, and there were a number of near misses and crashes in the wild West Coast conditions.

While some 'baiters' lived communally, sharing resources and working together, others clashed over the best fishing spots. In 1928 Din Nolan started up a large

ETYMOLOGY

The name *Galaxias* means 'galaxy', as the spots on the giant kōkopu were thought to resemble the Milky Way galaxy. The name 'īnanga' is derived from a Polynesian word for tiny fish or fish fry (immature juveniles) and is used both as a collective term for all whitebait, as well as the name for *Galaxias maculatus*. The name 'kōkopu' is also a Polynesian word and is used for freshwater fish species such as the Rarotongan brown gudgeon (*Eleotris fusca*). Whitebait is a common term in England for any small edible juvenile fish, especially sardines and herring. In New Zealand, adult whitebait were also known as 'native trout' or 'cow fish', as they can turn a river milky-white when they spawn.

whitebait canning factory at Okuru, south of Haast, and was described as ruling over rivers like a czar, stationing guards along the river and taking people's catches if they tried to fish his patch.

A West Coast whitebaiter checking his nets. *(iStock)*

OUT OF THE FRYING PAN

Today whitebaiting remains a treasured part of many New Zealanders' lives. All sorts of characters flock to West Coast rivers for the whitebait run, living out of ramshackle shelters and caravans for several months of the year. Whitebait fritters, made with eggs and flour, have become a world-famous Kiwi icon and are sold at fish shops, markets and food trucks around the country. Whitebait remains one of our most expensive fish catches, sometimes selling for more than $150 per kilogram.

But all is not well for our native whitebait, and four of the five species you might find in a whitebait fritter (īnanga, giant kōkopu, kōaro and shortjaw kōkopu) are now threatened with extinction. Scientists estimate that several of these species may become extinct in the next few years if nothing changes, and all of them in a decade or so. Conservationists argue that New Zealanders should avoid buying whitebait until it can be managed sustainably.

BIOLOGY

The migrating galaxiids lay their eggs in vegetation along riverbanks. When they hatch, the larvae are swept out to sea and live there for several months before returning to fresh water. Though they appear similar as juveniles, each species is quite different in its habits. Īnanga don't get much bigger than their juvenile stage and stick mostly to the coast; the giant kōkopu grows to over half a metre long and is the largest galaxiid in the world; banded kōkopu has beautiful striped patterns on its skin and is found in clean forest streams; the kōaro is an expert climber and can ascend near-vertical cliff faces – some have been known to reach rivers and lakes up to 1300 metres above sea level; and the shortjaw kōkopu is rare, secretive and seldom seen. Galaxiids mostly eat insects, and have sensors on the top of their heads to detect the vibrations of prey that fall into the water from overhanging branches.

Whitebait runs have declined in recent years, and while there still can be decent numbers it is the adult fish that are most at risk. In many places, the riverbank vegetation they need to spawn has been removed, water quality has deteriorated, barriers prevent the fish moving upstream, and trout both eat them and compete for the same food. But there are promising signs on the horizon, with groups around the country protecting whitebait spawning sites, replanting native vegetation and building fish bridges and ladders for fish to pass over during their upstream migration.

Banded kōkopu were much admired by Māori for their beautiful markings. *(Shaun Lee)*

A group of banded kōkopu in the shelter of a shaded forest stream. *(SeacologyNZ)*

Upokororo / New Zealand grayling

The phantom fish

TAXONOMY

Upokororo (*Prototroctes oxyrhynchus*), or New Zealand grayling, is our only fish species known to have become extinct since the arrival of humans. It belongs in the smelt family, Retropinnidae, which includes the porōhe or New Zealand smelt (*Retropinna retropinna*). Its closest living relative is the Australian grayling (*Prototroctes maraena*), which is under threat from both disease and introduced trout.

While every New Zealander grows up learning about the long-lost moa, fewer tales are told of the upokororo. It was once one of our most common freshwater fish, so abundant that it was said a fisherman could 'kill till his arm tires and his basket will hold no more'.[10] But the last confirmed sighting of upokororo was around 100 years ago, and they have never been seen alive since.

Even in life, the fish was a bit of a phantom, and was described as disappearing and reappearing like a ghost. The slightest disturbance, or wayward shadow falling upon the water, would be enough to send upokororo scattering. To catch them, Māori fishers would look for areas where algae had been nibbled from the rocks. As soon as the fish were spotted, the fishers would immediately back away from the river so as not to startle them. A popular method to secure them was to split into two groups, with one group taking nets downstream while the other started beating the water with sticks to drive the fish into the waiting nets. It must have been a bitter disappointment when the fish took fright before the fishers could ready the nets; in one tradition the disappearance of the fish was blamed on a nefarious tohunga named Kaaho, who scared the fish away using his dark magic.

Upokororo were relied upon as an important seasonal food source, especially on the west coast of the South Island. The ideal time to catch them was in late summer, when they fed on freshwater algae and their flesh became extra fat. It was apparently incredibly rich, and if you ate it too fast you could make yourself sick. The catch was preserved by pulling out the entrails with a stick, cooking the fish over burning embers or in a hāngī, and then packing it into baskets for future use.

Upokororo (*Prototroctes oxyrhynchus*) by Frank Edward Clarke (1889). *(Te Papa, 1992-0035-2278/1)*

GOOD SPORT

Many European settlers had a strongly held belief that nothing of value lived in New Zealand rivers, but upokororo was a rare exception. It was considered a fine sporting fish, compared to the English grayling and regarded as a delicacy – eaten smoked, cured or fried in its own fat.

The upokororo was a favourite among sport fishers, who liked to fish for them in murky conditions when they were less likely to take fright, with the best bait said to be the little red worms that lived underneath cowpats. One hundred fish in an afternoon was considered a decent effort, and one fishermen reported catching over 500 in a single sweep of the net. Even acclimatisation societies loved this fish, and there was a coordinated effort to increase the numbers of upokororo in east coast rivers. Societies made deals, trading trout and perch for upokororo, and they were bred at the Masterton aquarium ponds, from where thousands of baby fish were sent out to nearby rivers.

THE FINAL DISAPPEARING ACT

At the height of their abundance upokororo were exploited ruthlessly. When they retreated to dark caves and crevices during the day, some fishermen took to dynamiting them, causing entire shoals to float to the surface, where they could be scooped up and collected. The introduction of trout no doubt took its toll as well, as upokororo had evolved in complete isolation for millions of years and likely didn't have the ability to cope with these new invaders. With so much of the landscape being converted from forest to farmland, many rivers may have become uninhabitable and the algae the fish relied on as food might have disappeared.

ETYMOLOGY

The name *Prototroctes* means 'first trout', while the species name *oxyrhynchus* means 'sharp snout'. The European name grayling was given by settlers who thought it looked similar to the northern-hemisphere grayling (*Thymallus thymallus*), although they are not closely related. The Māori name 'upokororo' roughly translates as 'brains in the head', and other names for this fish all refer to the brains or head, e.g. upukorokoro, upokokororo, pokororo and paneroro. This could mean that upokororo was considered a clever fish that was difficult to capture, or it may have referred to the fact that its brains were large and full of fat.

The last confirmed sighting of upokororo on the Waiapu River in 1923. *(Alexander Turnbull Library, 1/2-037936-F)*

By 1870 the first declines were being noted, by the 1880s upokororo had vanished from most rivers, and by the early 1900s sightings were rare. When shoals appeared it was enough to make the newspapers, and if a fish was caught it might be sent to a biologist to identify it or displayed in a shop window as a curiosity. In 1923, with the fish now increasingly rare, the anthropologist Sir Peter Buck (Te Rangi Hīroa) asked a group of Waiapu Māori on the North Island's east coast to show him their traditional methods for catching upokororo before it was too late. They built a trap out of supplejack vines and harakeke leaves and managed to catch around forty. This was the last ever confirmed sighting of the species; occasional unverified sightings have trickled through over the past century, but scientists believe the species is now well and truly extinct.

The Australian grayling (*Prototroctes maraena*) is the closest living relative of the upokororo. *(Greg Wallis)*

WHAT KILLED THE UPOKORORO?

BIOLOGY

Upokororo were once reported in almost all clear-running streams around New Zealand. They most likely spawned in fresh water, and the larvae were swept out to sea, where they joined the other whitebait species. It appears they did not possess the ability to climb, so they probably didn't reach too far inland. They were most likely herbivores, grazing on the algae growing on rocks, using their upper jaw full of comb-like teeth, but they would also bite at bugs and worms used as bait by fishermen. There are some reports that they retreated to darker caves and crevices during the day and then moved to shallower waters at night to feed.

For a long time this relatively recent extinction was a complete mystery. Many people blamed the introduction of trout, but no one could explain why the fish had also disappeared from rivers where there were no trout. A recent analysis may have solved the mystery. It showed that, after spawning, upokororo eggs were swept out to sea, but the juveniles, instead of returning to the rivers their parents lived in, may have swum into rivers at random.[11] This meant the fish would have faced a lottery: they could end up in a 'safe' river or they could end up in a degraded river full of trout, nets, fishing rods and dynamite. Even though there were still pockets of good habitat, over time these toxic rivers would have been common enough to slowly drive the species to extinction. It may have taken relatively few of these rivers to finish off the upokororo, as it seems to have been uniquely vulnerable. With no climbing ability, these fish would have struggled to escape predation from trout, and their shoaling behaviour meant they could be caught in large numbers. In addition, they were the only known freshwater species to rely on an algal diet and thus would have been massively impacted if this food source disappeared.

The fate of the ghostly upokororo is particularly relevant for our other threatened freshwater fish, such as the longfin eel, which travels out to sea to spawn and whose juveniles also return to rivers randomly.

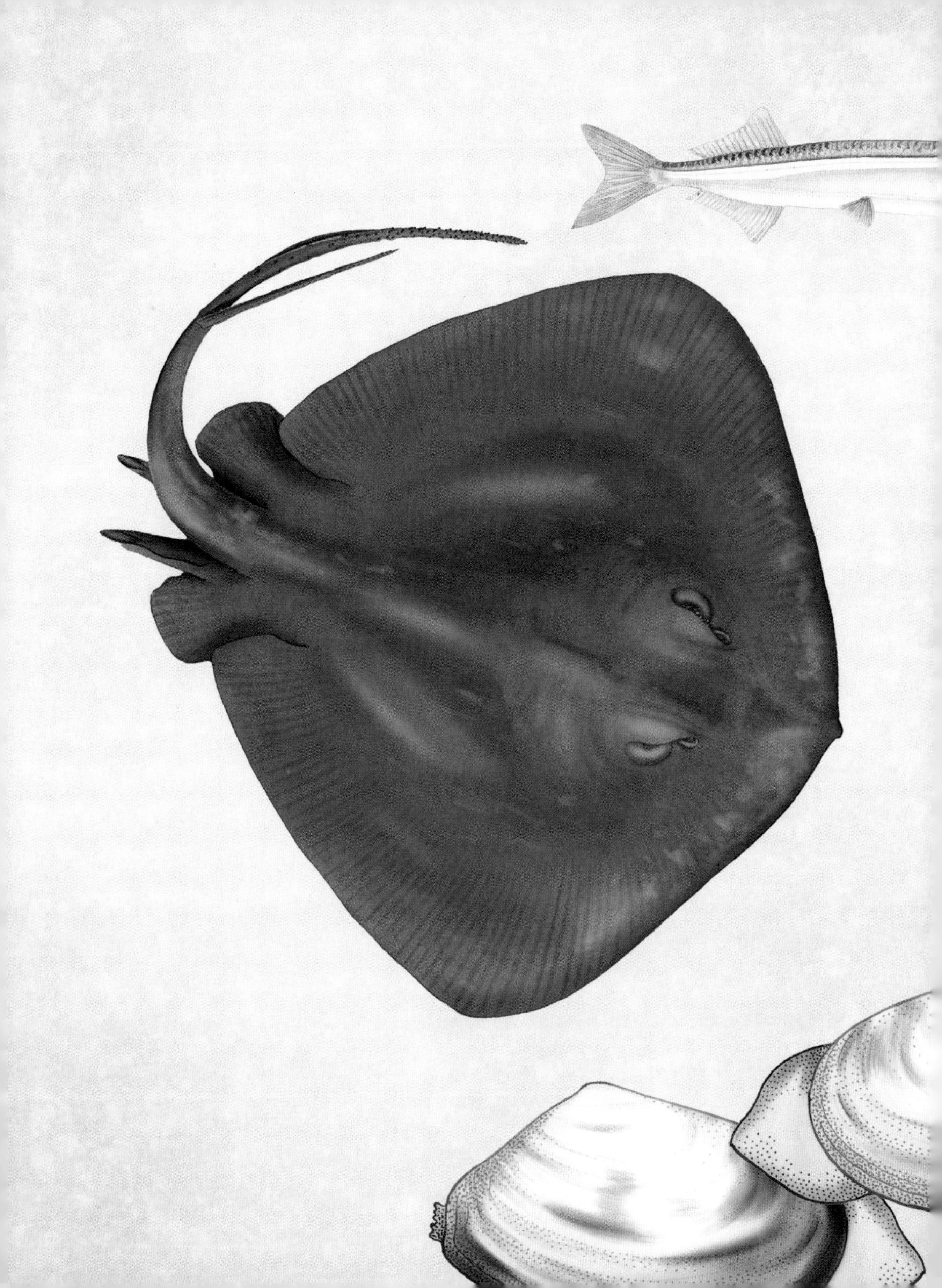

SANDY SHORES

Inhabitants of estuaries and beaches

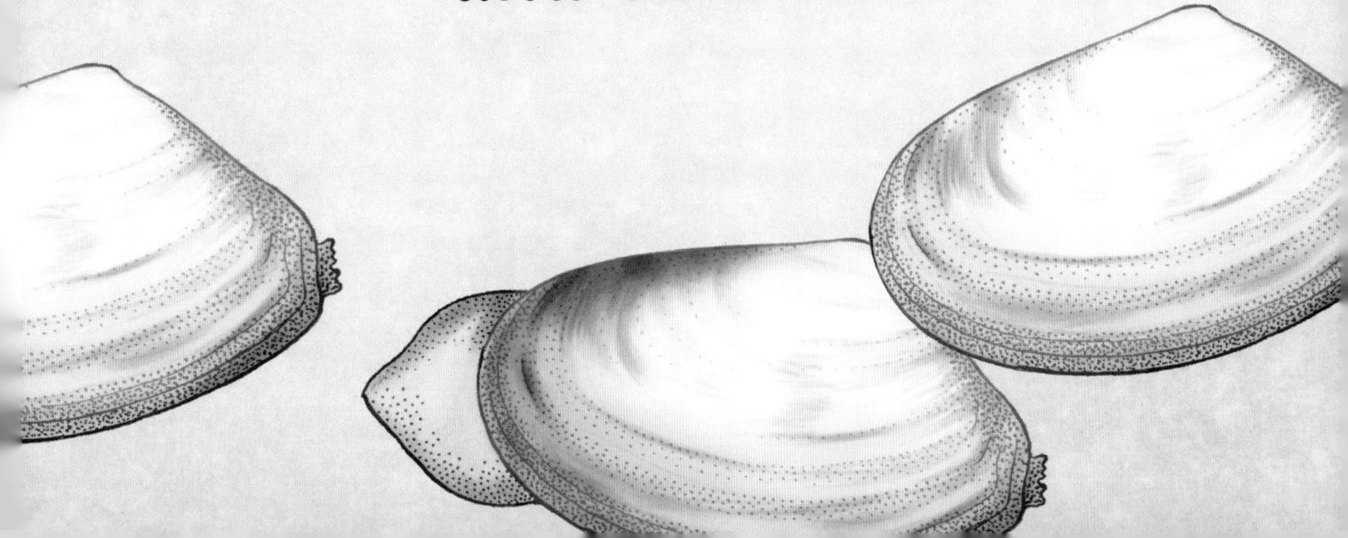

Cockles / Tuangi

Treasure of the tide

New Zealand cockles (*Austrovenus stutchburyi*) by Edgar Smith (c.1844), adapted by Lars Quickfall. *(Biodiversity Heritage Library)*

Cockle shells are so common, often seen scattered across the harbour sands at low tide, that they blend in as part of the landscape. It's only when they are submerged in water that cockles spring to life, waving their siphons about, sucking up water and catching bits of plant and animal matter.

Though they may be small, they can reach astounding densities, with thousands of cockles in a single square metre of sand. This makes them an attractive target for predators, and they are food for a wide range of species such as paddle crabs, flounder and seabirds. Research at the Avon-Heathcote Estuary/Ihutai in Christchurch estimated that a single South Island oystercatcher (*Haematopus finschi*) eats around 200,000 cockles per year, with a flock of oystercatchers eating more than 400 million over the course of a year.[1]

A USEFUL HARVEST

This huge abundance of cockles also made them a valuable food source for Māori, who know them as tuangi or tuaki. No one could go hungry when cockles were around, as they were easy to access and available for harvest year-round. They could be simply cracked open and eaten raw, but the preferred way to eat them was to cook them just long enough for the shell to begin to open but before they became too rubbery.

Typically, women led the harvest of cockles, digging in the soft sand and mud with their feet and manoeuvring the shellfish into a woven kete or flax bag. These kete were ingeniously designed so that only cockles of the right size were kept, while those that were too small slipped through gaps in the weave.

The sandy bottoms of harbours and estuaries are often littered with empty cockle shells. *(John Barkla)*

Empty cockle shells provided a useful resource as well. They could be used to pluck out stray hairs when trimming facial hair, and the rough side could be used for peeling vegetables or scraping flax to prepare it for weaving. They could be given a sharp edge and were used to lacerate the skin as a sign of mourning after the death of a loved one.

Because cockles are so easy to collect there was a risk of overharvesting, so Māori carefully managed cockle beds like a community garden. At times when the numbers seemed to drop, closed seasons and rāhui were placed on the beds to ensure numbers could build up again. Particularly good cockle beds were jealously guarded and defended from others. Otago Harbour was well known for its large cockles, and the harbour was divided up and marked out with wooden stakes, ensuring only those with whānau and hapū rights could gather there.

TAXONOMY

Cockles or tuangi (*Austrovenus stutchburyi*) are endemic to Aotearoa. They are not true cockles (Family: Cardiidae) like the common cockle of northern Europe, but are part of the large Venus clam family Veneridae, which includes many of the shells washed up on New Zealand beaches, such as the ringed dosinia (*Dosinia anus*).

Breaking cockle protocols could have serious implications. In one tradition from the early seventeenth century, after a school of orca beached in Golden Bay

Cockles were an important food source that was easy to collect and available year-round. *(iStock)*

a tohunga made the nearby cockle beds tapu and prohibited their use. But some members of the local Ngāi Tara iwi harvested cockles from the beds. Ngāi Tara were later badly defeated and driven out of the area by Ngāti Tūmatakōkiri, and some believed this was divine retribution for breaking the tapu.

SILVER BELLS AND COCKLE SHELLS

Cockles found a place in the heart of Pākehā and many have fond memories of collecting them from the shore. They were sold in a few places, but were more important as a cheap, readily available wild food that could be harvested at any time. A quick run to the cockle beds in the morning could supply breakfast for a family, and they were typically eaten simply, either by themselves or with bread, or could be cooked into larger meals, sometimes served with white sauce and butter.

Cockles were formerly abundant on the harbour beaches of Auckland, and it was a favourite pastime among local rugby players to collect cockles on Sunday morning while reminiscing about the game the night before. Some farmers found a use for the shells, grinding them up for chicken feed, and the shells were sometimes covered with silver or gold and used as earrings, buttons and belt buckles.

THE MASTER MANIPULATOR

As such an abundant feature of estuary ecosystems, cockles are not just the target of predators but of parasites as well. One devious parasite, the trematode flatworm (*Curtuteria australis*), uses cockles as pawns in its grand master-plan to infect oystercatchers.

The flatworm infects the body of a whelk (a type of marine snail) and begins producing large numbers of larvae. These larvae are then released into estuary waters to be sucked up by the cockles as they feed. Once inside a cockle, the flatworm attacks the muscle that it uses to dig itself into the sand. The muscle wastes away, leaving the cockle helplessly stranded on the shore when the tide recedes.

With no way to escape, such cockles are easy prey for oystercatchers, which eat both cockle meat and flatworm. Once inside the oystercatcher, the flatworms complete their life-cycle by laying their eggs in the bird's stomach. The eggs are passed out in the bird's faeces, where they are eaten by whelks and hatch to begin the cycle again.

ETYMOLOGY

The name *Austrovenus* means 'southern clam', while the name *stutchburyi* was given in honour of the nineteenth-century geologist and biologist Samuel Stutchbury. The Māori name 'tuangi' (pronounced tuaki in the south) comes from the Polynesian word 'tuahi', which is used for other clams and originally referred to a tool for grating coconuts.

The English word 'cockle' may come from the French *coquille*, meaning 'shell' or 'scallop', or from a corruption of the Latin word *cochleae*, meaning 'ventricles of the heart'. When viewed in profile, a cockle resembles the shape of a heart, and there is a long connection between the two words, as in the phrase 'warming the cockles of your heart'.

A pied oystercatcher feeding on cockles, by George Edwards, 1790. *(Alexander Turnbull Library, A-191-031)*

Unfortunately for the poor cockles, these are not the only organisms that take advantage of them – a whole host of parasites hitch-hike on infected cockles to get a free ride to the surface. Some will attack and weaken the mantle around the edge of the cockle shell, effectively leaving the door open for more parasites to invade. However, cockles do have an ally in their battle against parasites – the mud anemone (*Anthopleura aureoradiata*). Mud anemones grow on the outside of cockle shells and eat up parasitic larvae, meaning both cockle and mud anemone are less likely to end up stranded on the surface.

Mud anemones (*Anthopleura aureoradiata*) colonising the shell of a cockle. *(Shaun Lee)*

BIOLOGY

Cockles are found in the intertidal sediments of estuaries and harbours throughout New Zealand, especially in soft mud and sand. If they are not eaten by birds, fish or people, they can live for up to twenty years. To age a cockle you can count the number of growth rings on its shell, like the rings of a tree stump, with the youngest layers closest to the hinge. The daily rhythm of the tide rushing in and out of harbours is so hard-wired into the cockle's biological clock that if they are removed from the shore and placed in a tank of sea water they continue to mimic their natural feeding patterns.

While the parasitic flatworm might be bad news for the cockle, it is believed to play an important role in estuary ecosystems. Because so many cockles are prevented from burrowing down into the sand, harbour floors become covered with a hard surface of cockle shells. Other marine organisms such as limpets, crustaceans, barnacles, marine worms and seaweeds all grow on these stranded shells, creating a richer and more diverse environment in areas where the parasite is most active.

Piper / Takeke

The fish with a spear on its face

TAXONOMY

New Zealand piper (*Hyporhamphus ihi*) is found only in Aotearoa. It has close relatives across the Tasman, such as the southern garfish (*H. melanochir*), and belongs to the halfbeak family Hemiramphidae. The family is also related to flying fish (Exocoetidae), and they share a remarkable ability to leap out of the water.

Despite being found all around New Zealand's coastline, piper can often be hiding in plain sight. They have developed a remarkable strategy for evading detection, hugging so close to the surface of the water with their long, silvery bodies that they are rendered almost invisible. It makes them very hard to spot, whether by fish from below or by seabirds from above. What's more, they can change their colour at will, by expanding or reducing pigment in their skin to make sure they blend in at all times. It is such a successful strategy that piper rarely venture much deeper than a few metres from the sea surface.

When they are spotted, piper have one last trick up their sleeves – leaping clean out of the water and soaring through the air in an attempt to throw off their pursuers. However, despite their best efforts, they are still a favourite food of a number of predators, such as dolphins, which use echolocation to track them down, and gannets, which plunge on them from great heights, taking them unawares.

Piper (*Hyporhamphus ihi*) by Frank Edward Clarke, 1875. *(Te Papa, 1992-0035-2278/40)*

LONG CHINS

Perhaps the most intriguing feature of these fish is what looks like a long jaw, but is actually an extra-long chin, reinforced with bony plates and coloured orange at the tip. Over the years a lot of theories were proposed for these awkwardly long chins: it was believed that they were used to spear smaller fish, used to dig for food in the sand, or even that they were swung like a scythe to slash down beds of seagrass to eat. In one Māori tradition, these strange chins were said to be the result of a great battle between fish and humans. The victorious fish were allowed to choose a prize from the spoils of war, and the piper selected one of the humans' spears and placed it on his nose.

Research has revealed that these chins are actually highly developed sense organs, studded with pores that can sense the movement of small animals floating in the water column. When the sun sets, piper leave coastal waters and move into estuaries and sheltered harbours to feed. Their eyesight is not adapted for hunting prey at night, so this is where their long chin comes in handy. They use it as a detection device, hunting out small prey swimming on the surface of the water, then scooping them up with their mouths.

One Māori tradition explains that the piper's long chin was a spear gifted by Tangaroa. *(Luke Colmer)*

THE FLIGHT OF THE TAKEKE

Until recently, piper were not considered to be an important part of the historic Māori diet. That was mainly because their small and delicate bones were difficult to detect using old archaeological methods. When midden sites were resampled with finer nets and more sensitive methods, piper bones turned up frequently, indicating they were probably a very common food source. This makes a lot more sense, as they are an abundant and easily caught fish.

Māori usually trapped them by dragging finely woven nets over seagrass beds in shallow harbours and estuaries. Great care had to be taken when scooping piper into the net, as they could easily take fright and disappear in every direction – although sometimes in their mad scramble to safety they were known to leap into the boats of the fishermen by mistake. Māori admired the ability of the little piper to leap from the water when avoiding the net, and someone who had travelled a great distance was compared to a flying takeke.

PIPER DOUGHNUTS

Piper quickly caught the attention of Pākehā fishers, who admired their sweetness and flavour. Biologist David Graham raved about them in the 1950s, writing that piper 'contains a flavour of gastronomic quality not found in other fish … and when used as a cold dish nothing is more palatable to the epicure'.[2] They often fetched a high price when sold at market, and there was enough interest in them that experiments were made on how to can them.

Huge shoals were reported in the past: one at Waihī Beach in 1929 was said to measure more than 3 kilometres long. The harbours of Auckland in particular were popular piper fishing spots, and some fishers could bring in hauls of over a thousand or more to sell at the fish markets. In 1890 so many piper were caught and smoked in central Tauranga that it was said that the surrounding neighbourhood smelled like smoked piper for some time after. Akaroa resident Bethia Latter recalled that in the 1940s the storekeeper who lived at the head of the bay would keep a watch out for piper; once a shoal was spotted, he would ring around all the homes and locals would rush out with nets and old petrol tins to collect them.

Today piper are often considered an 'entry-level fish' and many children enjoy catching them off a wharf, drawing them in with handfuls of breadcrumbs or Weet-Bix as bait. Piper are vicious fighters for their small size, and when one is caught on a rod and line it will leap out of the water like a miniature marlin.

ETYMOLOGY

The name *Hyporhamphus* means 'under beak', and *ihi* is based on one of its Māori names, ihe. The name 'ihe' and similar words (ise, sise) are used for this fish family across Polynesia, and originally referred to a dart or spear. The name 'takeke' is used more commonly in the North Island; its origins are unclear, but another similar Māori word, 'tākeke', means 'to be entangled in a net'. The word 'piper' references its long shape, like a pipe, and the name 'garfish' probably comes from the old English word *gar*, meaning 'spear'.

Many fishers catch piper in estuaries and use them as live bait for fish such as snapper, kingfish and kahawai. At night, piper are easy to spot, as a bright light strips away their cloak of invisibility and reveals their electric-blue bodies, which appear as if they are flying through the water. They are very easy to catch at night and become very docile, bumping into fishers' legs as they happily feed on the surface in the shallows.

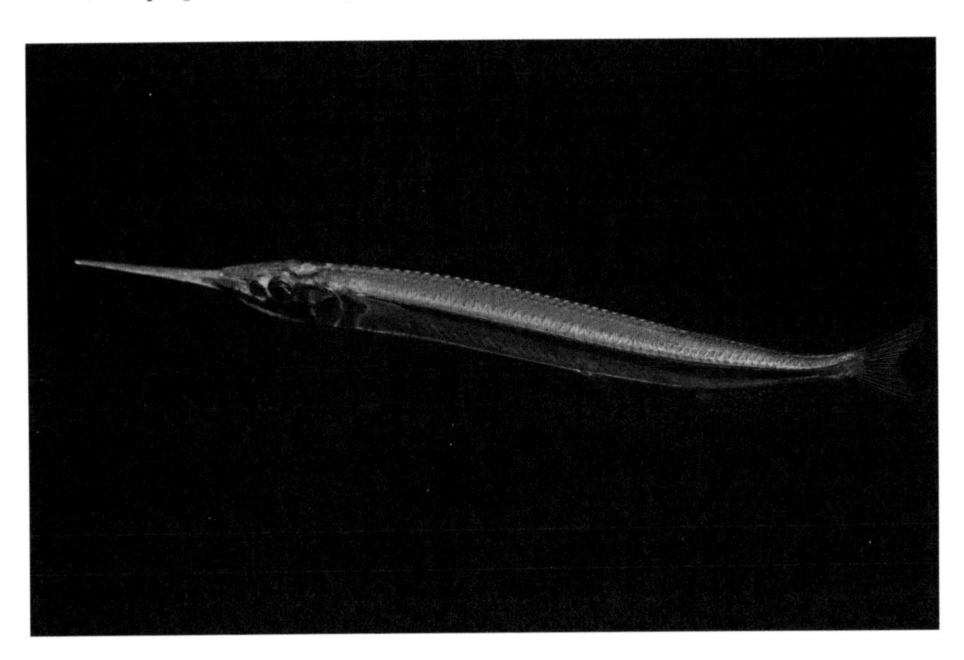

The striking blue colour of the piper is revealed under torchlight. *(Daan Hoffman)*

BIOLOGY

Piper are found all around New Zealand in sheltered harbours and reefs close to shore. They primarily feed at night on animal plankton such as mysids, crab and worm larvae, ostracods, copepods and cumaceans. They have tiny teeth and sometimes feed on eel grass (*Zostera* spp.) and sea lettuce (*Ulva* spp.). Their life-cycle is entirely dependent on the availability of estuaries and sheltered harbours in which to breed, as they lay long, sticky eggs that need to be attached to seagrass to survive.

Piper are still regarded as one of our tastiest fish, with the classic preparation being the 'piper doughnut'. The fish are rolled with a bottle to break the bones, and the spine is removed in one piece. Then the bill is poked through the tail to form a doughnut shape, and the fish are covered in flour and fried.

Toheroa

The long tongues

Toheroa were once our most famous shellfish, eaten in the dining rooms of Sydney and London, and described as 'a gift of nature … that has done much to advertise the Dominion all over the world'.[3] Found in immense beds several kilometres long, so dense they could be harvested with a horse and plough, toheroa were believed to be an inexhaustible resource. But today their populations are so threatened that they are the only shellfish that it is illegal to take or disturb without a permit.

For Māori, toheroa have always been highly significant. To make the most of the huge abundance of food they provided, communities would set up temporary camps near the coast to collect and preserve toheroa for future use. The shellfish were dug out by hand and placed in woven kete that allowed the sand and water to drain out. Once gathered, toheroa could be cracked open and eaten raw on the beach, cooked in a hāngī or preserved by threading them with strings of flax and leaving them in the sun to dry.

Even a single toheroa is a rich and sustaining meal, and a ration of one toheroa a day was fed to taua (war parties) on the march. Desirable toheroa beds were themselves sometimes a cause of war, and there are accounts of battles fought over the large beds on the west coast of the North Island.

Those who had access to toheroa beds were in possession of a valuable resource. Dried toheroa were useful items for trade, and were seen as a prestigious food to serve to guests when welcoming them onto the marae. Indeed, toheroa were valued so highly that Māori actively translocated them across the country to introduce them into new areas and build up population numbers. Oral traditions speak of toheroa being stored in bags of bull kelp filled with sea water to keep them alive on long journeys. This may help explain why toheroa have such an unusual distribution: primarily along the west coast of the North Island, but with one isolated population at the bottom of the South Island.

THE STORY OF PĪNGAO

As such an important shellfish, toheroa were said to have very prestigious origins. They were believed to have originally come from the spiritual homeland of Hawaiki and been 'planted' on the west coast of Northland. Another tradition from the north helps explain the biology of toheroa by describing its special relationship with pīngao, a golden sand sedge. Pīngao was originally a seaweed, but was placed in the dunes by Tangaroa – atua of the sea – to look after toheroa.

Toheroa (*Paphies ventricosa*) by Arthur Powell (1947), adapted by Lars Quickfall. *(Auckland War Memorial Museum, Tāmaki Paenga Hira)*

It is believed that at the spring tide, the tiny juveniles of toheroa (spat) are carried in the foam of the waves and deposited within the leaves of pīngao on the sand dunes. Here they are nurtured and grow strong until they are ready to survive in the sand. Then they hitch a ride on the seed-heads of spinifex grass – the tumbleweed-like plants that blow across sandy beaches – and are deposited across the beach. When an expert in toheroa deemed this process was taking place, children were forbidden to play waiwatai, a game in which they chased the tumbling seed-heads around the beach, in case they disturbed the toheroa during this vulnerable time.

ROYAL GREEN SOUP

Pākehā enjoyed eating toheroa from the early days of settlement. The shellfish were left in a bucket of water to encourage them to spit out any sand and then cooked over a fire, fried, steamed, minced or turned into fritters and served with bread. For the purists, however, there was only one way to eat them: toheroa soup.

At first glance toheroa soup seems an unlikely food phenomenon, as when the shellfish are boiled it results in an unusual, khaki-green broth, coloured by all of the phytoplankton they eat. But this green soup was highly admired. It had a delicate flavour that was easy on the stomach but was nourishing and invigorating at the same time.[4]

In the 1920s, toheroa soup went global. During a royal tour of New Zealand, Prince Edward (later King Edward VIII) was served the soup at a banquet and loved the taste so much that he underlined it on the menu and scribbled in the margin 'Very good!'[5] The incident was reported widely in newspapers, spurring huge demand across the British Commonwealth, as people sought to try this dish 'fit for a king'. Toheroa soup was soon

TAXONOMY

Toheroa (*Paphies ventricosa*) is a surf clam in the family Mesodematidae. *Paphies* is found only in New Zealand, and also includes other popular edible shellfish such as pipi (*P. australis*), northern tuatua (*P. subtriangulata*) and southern tuatua (*P. donacina*).

Three girls shelling toheroa on a beach in the early twentieth century. *(Alexander Turnbull Library, 1/1-026522-G)*

on the menu at restaurants and diners throughout New Zealand, Australia and Britain. It became a staple of high-end restaurants as well as milk bars on the street, and hot toheroa soup was a favourite in the cold English winter.

In response to the demand, the toheroa-canning industry grew substantially and huge numbers of the shellfish were canned whole or minced and sent overseas. At the height of toheroa's popularity, Australian fisheries managers investigated how they could introduce the shellfish to Australia, and folk tales rose up about a wealthy American businessman who attempted to buy New Zealand solely to take possession of the supply of toheroa soup. Toheroa collecting became something of a national pastime, and held such a prominent place in the public consciousness that children sang about toheroa at school and canned toheroa were sent to New Zealand troops fighting in World War II in Egypt and the Pacific.

TOHEROA WARS

But even as early as the 1920s there were concerns that toheroa were declining so fast they might become extinct if nothing was done to protect them. In the 1950s and 1960s, with more New Zealanders having access to cars, it became popular to drive to the beach to collect toheroa. In 1966, at Glinks Gully at Ripirō Beach in Northland, around 50,000 people turned up and harvested over a million toheroa in one weekend.

Toheroa digging at Muriwai Beach, 1962. *(Archives New Zealand, R24459648)*

Over time the huge beds began to shrink, and fewer and fewer toheroa were found every year. By the end of the 1960s, the toheroa canning factories had closed down, and tight restrictions were being introduced, with harvesting only allowed on occasional open days. These days could be wild and frantic affairs, as thousands of people descended on the coast all at once to collect toheroa. Cars sometimes collided on beaches and conflicts broke out between gatherers. Some people got so caught up in the excitement they didn't actually know what to do with the shellfish they caught, and surplus toheroa could sometimes be found at the local dump afterwards.

ETYMOLOGY

The name *ventricosa* refers to the valves on the shell, which curve inwards. The Māori name 'toheroa' means 'long tongue', and is the common name in English as well.

Another explanation for the name comes from a Māori tradition that a group of bird hunters, running away from their enemies on Te Oneroa-a-Tōhē/Ninety Mile Beach, were desperate for food and began digging in the sand. They were about to give up when they heard the message 'tohe roa, tohe roa', which means 'carry on and persist'. They finally discovered the shellfish which saved their lives, and they gave it the name 'toheroa'.

Toheroa are surprisingly fast and active diggers, using their muscular foot to dig themselves into the sand.
(Lloyd Esler)

Toheroa hunters look for siphon holes in the sand, which often indicate the presence of a toheroa bed. *(Lloyd Esler)*

The last open day was in 1993, and ever since toheroa have been completely protected, with only small numbers allowed to be taken for the customary Māori harvest. Today, toheroa poachers risk fines of up to $20,000, yet they often go to extreme lengths to steal the shellfish. Some have been caught stowing toheroa in handbags, jacket pockets and spare tyres; pretending to play golf near the beach and digging them up with clubs; or getting their children to build sandcastles over the beds so they can be collected in secret.[6]

STARTING AGAIN

Even after forty years of restrictions, toheroa numbers have failed to recover. They remain under threat from a range of sources, such as disease, run-off, pollution and vehicles driving on beaches. However, there is some hope for the future, as a female toheroa may release 15–20 million eggs in a single spawning event. If some of the threats to toheroa could be managed, then their populations have the potential to rebound. Work has been done to investigate restoring toheroa beds, with kaitiaki Māori experimenting with traditional methods of translocating toheroa using bull kelp bags to help build up populations. In one example, a group of kaitiaki took 30,000 toheroa from Ripirō Beach and moved them to Te Oneone Rangatira Beach, near Muriwai. They have successfully established there, creating a model for how other toheroa beds might be restored in future.

BIOLOGY

Toheroa are filter feeders, consuming the rich phytoplankton which is abundant at high-energy surf beaches. While most toheroa stay in one place, some individuals will travel several kilometres to other beds. Occasionally entire beds of toheroa will pick themselves up and move away, releasing their hold on the sand and allowing the waves to move them across the beach. Whole beds of toheroa have been observed moving in this way, sometimes 30 metres or more in a single night. On black-sand beaches on the west coast their shells are sometimes stained a blue-black colour from the iron content of the sand.

Paddle crab / Pāpaka

Cannibals with claws

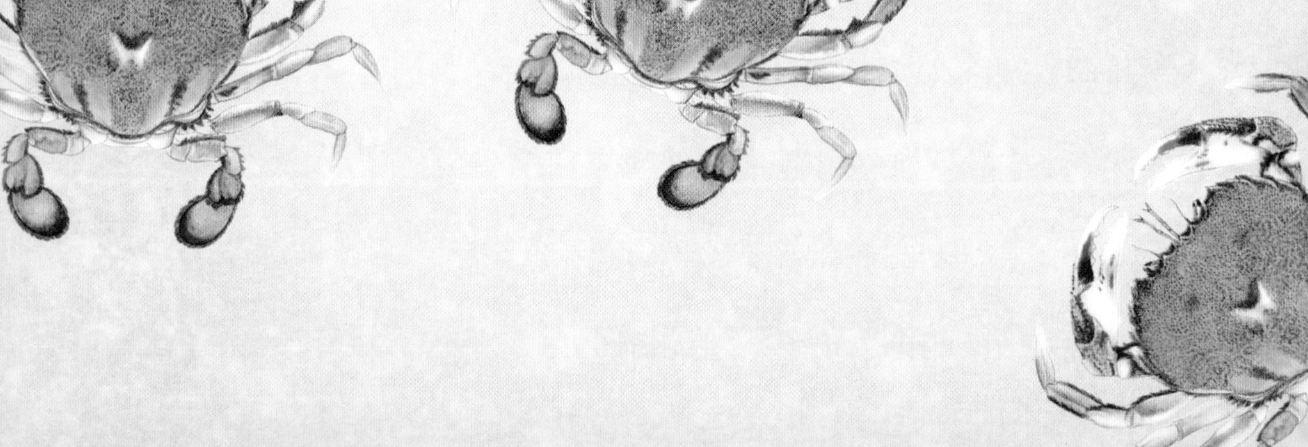

Paddle crab (*Ovalipes catharus*) by Kawahara Keiga (c.1820s), adapted by Lars Quickfall.

While most crabs rely on being able to scuttle into rocky crevices to hide from predators, paddle crabs are masters of the sandy shore, surfing through the currents with their oar-like back legs. If they need to disappear from a predator, they can burrow into the sand, disappearing in an instant, save for two beady eyes sticking up. This digging ability comes in handy when they are hunting for buried shellfish, which they excavate with their strong legs and pull apart with their powerful claws.

Paddle crabs can disappear below the sand in an instant. *(Daan Hoffman)*

FAIR GAME

Paddle crabs aren't choosy when it comes to finding food on the sandy shore, and anything is fair game – even other paddle crabs. In fact, other paddle crabs can make up as much as a third of their diet in some places. This means that paddle crabs can never really trust their neighbours, and have to be especially careful when they have outgrown their hard exoskeleton and need to moult and begin growing a new shell. A large paddle crab will think nothing of feeding on a defenceless soft-shell crab, and juveniles often have to hide from larger crabs while they are moulting.

Paddle crabs are so shameless when it comes to food that they have even been known to seize the eggs of females as they are being laid. A laboratory study observed small paddle crabs circling a female, cutting off pieces of her egg mass with their claws as it was extruded, and eating it.

A KNIGHT IN SHINING ARMOUR

When it comes time to mate, paddle crabs will court one another with 'rasps', rubbing their legs together to make a scratching sound in the same way that cicadas do. Females can only mate once they have moulted, so males will often go to great lengths to protect their mates during this delicate time. The female tucks herself into a small ball and is held underneath the male by a pair of his walking legs. The male then continues as normal, swimming, burrowing and feeding, with the female tucked underneath him. It can take females anywhere from three to seventeen days to moult, during which time the male will vigorously defend her from any attack, making loud aggressive noises to deter other males.

TAXONOMY

Pāpaka (*Ovalipes catharus*) is the most common New Zealand paddle crab species, and is native to both New Zealand and Australia. It is a crustacean in the order Decapoda, which also includes crayfish (*Jasus edwardsii*). If you compare a crab and a crayfish they have the same basic body plan, but in the crab the tail is folded underneath to form a squat, compact body.

Once the female is ready, mating can take anywhere from twelve hours to four days. The male continues to guard her afterwards, while she is still vulnerable. Thankfully for the females, instances of 'post-coital cannibalism' are rare and males typically avoid eating females they have mated with. In some cases, large male crabs have been observed misunderstanding signals, picking up small male crabs that are about to moult and carrying them around and defending them by mistake.

Despite the potential danger mating holds for females, they typically have far higher survival rates than males, which are particularly vulnerable when soft-shelled, as they have no one to guard them.

A male paddle crab guarding a female during mating. *(Luke Colmer)*

A close-up of the sharp pointed carapace of a paddle crab. *(Daan Hoffman)*

SYMBOL OF UNITY

The distinctive shape of the paddle crab or pāpaka made it a prominent feature in Māori art and culture. It occurs as a weaving pattern in whāriki (woven mats) and is an inspiration for tā moko. It also served as a model for carvings in meeting houses and whare wānanga – houses of learning, where tohunga would pass on expert skills – and was adopted as a symbol of unity, whanaungatanga and tribal alliances.

ETYMOLOGY

The name *Ovalipes* means 'oval footed', and *catharus*, which means 'immaculate', probably refers to the shell. The English name 'paddle crab' relates to the creature's ability to swim through the water using its paddle-like hind legs.

Pāpaka is the Māori name for this species of crab but is also a general term for crabs, such as pāpaka huruhuru (hairy crabs) and pāpaka whero (red crabs) and even some insects such as pāpaka nguturoa for weevils (literally, the 'crab with the long beak'). Te Pāpaka-a-Māui is the name for Australia, but it is not clear why this name was chosen, as the landmass doesn't look much like a crab at all. It was probably first used when Māori began sailing to Australia in the eighteenth century to trade flax and other produce.

The Māori name for paddle crab, and crabs in general – pāpaka – features in the names of a number of pā sites and hapū around the country.

Despite the cultural impact of paddle crabs, however, surprisingly little has been recorded regarding their harvesting by Māori, but they would surely have been eaten historically, given they are widely consumed in Polynesia. The lack of information compiled by nineteenth-century ethnographers might suggest crabs weren't a key target for fishing, but they were probably collected opportunistically while Māori were harvesting pipi, tuatua and toheroa, or occasionally swept up in nets in shallow harbours and estuaries.

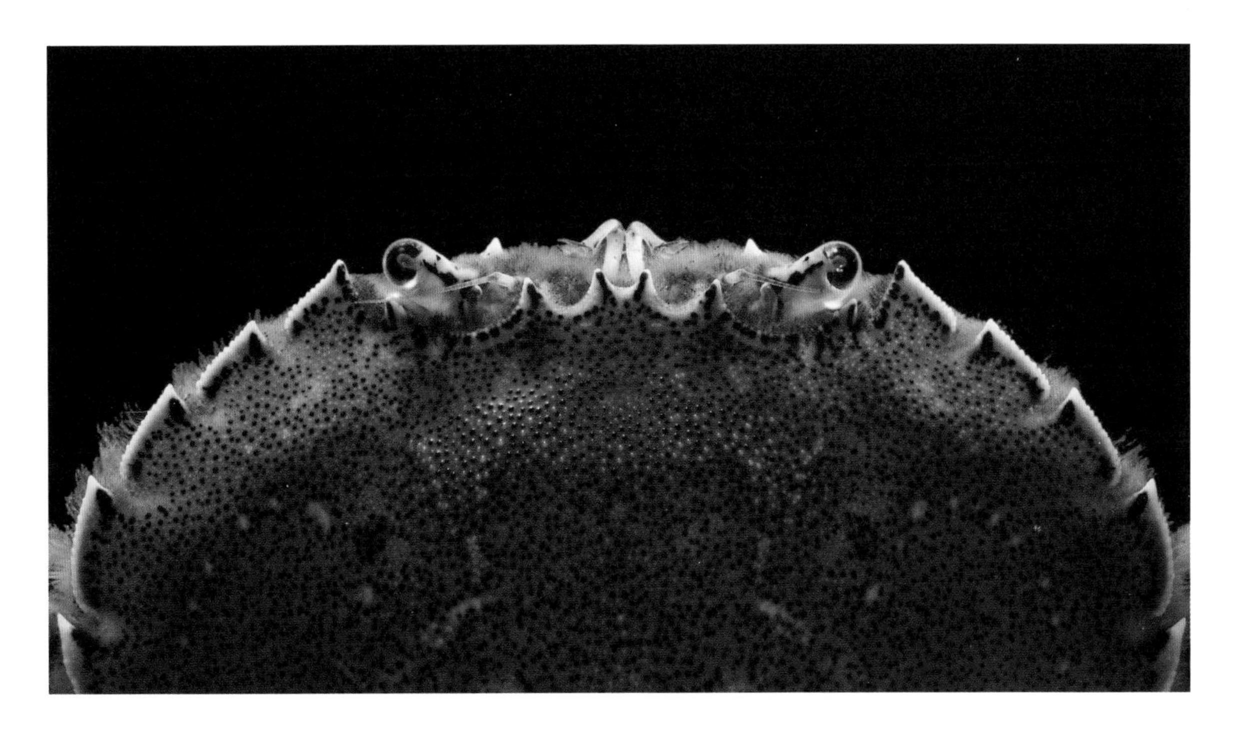

A NIP AT THE TOES

Paddle crabs were overlooked as food by many early Pākehā, as they were considered awkward to prepare and eat, and not particularly tasty. For those who were in on the secret, however, they were highly admired and their meat was deemed more delicate than that of crayfish.

Paddle crabs can be fished for with string and bait, but are more often caught in a baited pot on sandy beaches. Their aggressive nature means that fishers need to be quick to retrieve their pots, as the crabs will soon start fighting each other and tearing off limbs.

Anecdotally, populations of paddle crabs have increased in some areas, which suggests signs of an ecosystem out of balance. It's likely that the natural predators of paddle crabs – snapper, school sharks and elephant fish – have been overfished in many areas, allowing paddle crab populations to grow.

For most people, their main interaction with paddle crabs is when the crabs nip their toes at the beach. As noted, they are aggressive and will often put up a fighting stance rather than run away.

BIOLOGY

Paddle crabs are found between the high- and low-tide mark all around New Zealand and southern Australia, but can also be found in water as deep as 100 metres. They are incredibly effective predators of shellfish and specialise in eating pipi, tuatua and toheroa. Only very large crabs can get into a toheroa. First they cut off the siphon, forcing the shellfish to come to the surface of the sand, then they proceed to slowly chip away at its shell with their claws and hammer it on the sand to crack it open. The recent arrival of the Asian paddle crab (*Charybdis japonica*) in New Zealand waters is a concern, as it is much more aggressive, outcompetes the native paddle crab and can eat large numbers of native shellfish.

The oar-like hind legs are used to propel the crab through the water. *(Daan Hoffman)*

Paddle crabs are aggressive and rarely back down from a fight. *(Daan Hoffman)*

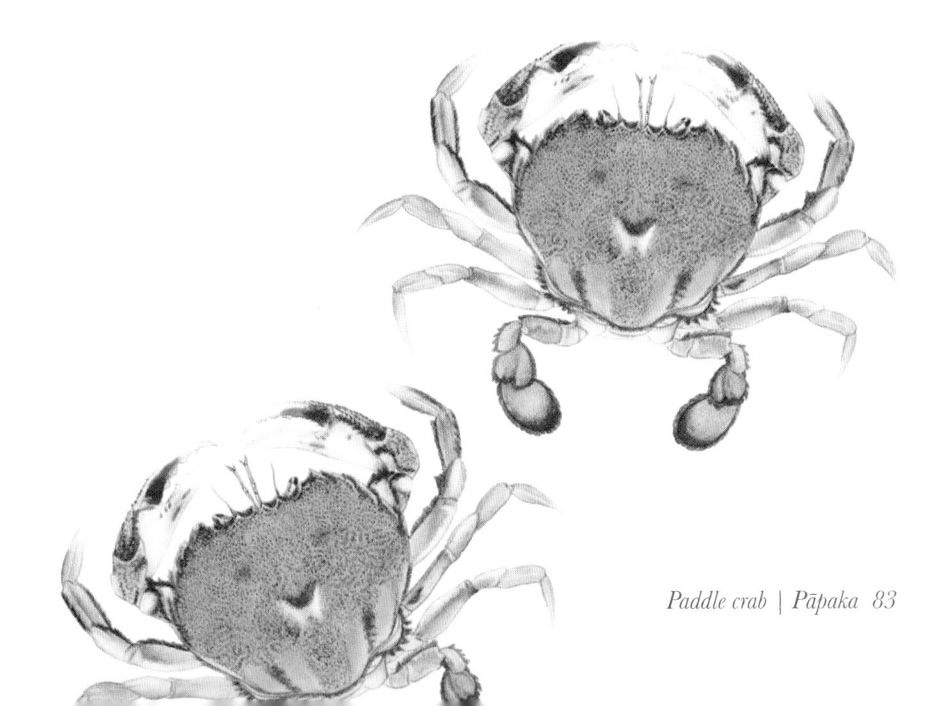

Flounder / Pātiki

Chameleons of the sea

Sand flounder (*Rhombosolea plebeia*) by Frank Edward Clarke, 1873. *(Te Papa, 1992-0035-2278/26)*

Flounder undergo one of the most remarkable transformations in the animal kingdom. They begin their life in the same symmetrical shape as any other fish, but as they grow, their left eye migrates from one side of their head to the other, and their jaw is slowly twisted around into a bizarre, lopsided smile. For a while they attempt to swim about awkwardly, but eventually they become exhausted and find a resting place on the sea floor.

The flounder's strange squashed face has been compared to a Picasso painting. *(Ian Skipworth)*

THE SQUASHED FISH

Māori recognised the unusual shape of flounder and believed it was caused by Hine-te-iwaiwa, the atua of childbirth. As she was chasing after the handsome Tinirau, she asked flounder for help. Flounder refused, so she squashed him flat in revenge. Another tradition says that flounder was sleeping in the house of Tangaroa at the bottom of the sea when Ruatepukepuke arrived to seek revenge for the death of his son. He set Tangaroa's house on fire, attacking each of the fish family inside as they emerged. Flounder was hit in the face, his jaws were crushed to one side and his eye knocked to the other side of his head.

These strange-looking fish were a valued and reliable food source for Māori. They were most often caught by dragging nets across a harbour during the day, or by taking a flaming torch of

TAXONOMY

The best-known species of flounder in New Zealand are the sand flounder (*Rhombosolea plebeia*), with its distinctive diamond shape, and the yellowbelly flounder (*R. leporina*), with its yellow underside. The black flounder (*R. retiaria*) has a black back and orange spots and is the only species that can live completely in fresh water, with some recorded in rivers up to 100 kilometres inland.

rimu or kahikatea wood and wading in estuaries at night, spotting the flounder by the torchlight and spearing them with a sharpened stick. If the spearer missed, the flounder would dart away, leaving a trail of sand in its wake. This led to the whakataukī 'The flounder will not return to the mud it has stirred up', meaning that someone who has caused trouble doesn't hang around to suffer the consequences.

ETYMOLOGY

The name *Rhombosolea* is a combination of words meaning 'diamond-shaped flat fish'. The species name *plebeia* means 'commonplace', and *leporina* means 'rabbit-like', possibly a reference to its overbite. The word 'pātiki' is a Polynesian word for flounder; in Te Reo Māori it can also be used as an adjective to describe anything flat. The English name 'flounder' relates to the way the fish flap and splash about in shallow water. They are also known as 'tin plates', 'three-corners' and 'flatfish'.

Different species of flounder were recognised and fished for at different times of year, such as the mohoao (*Rhombosolea retiarii*), which could be found in estuaries and freshwater rivers and was valued for its fatty flesh. Once flounder were caught, they could be cleaned, beheaded, cooked in a hāngī or hung up to dry for future use. When needed for food they could be pounded to soften the flesh and roasted, or grilled over the fire. Iwi and hapū who lived near large, shallow harbours were particularly fortunate, as they could rely on a steady supply of flounder year-round. Te Waihora/Lake Ellesmere near Christchurch was famous for its flounder, and when the bar would break periodically the tide would rush in and the lake become filled with these fish. All of the species of flounder could be found there, and at times they were so abundant they were said to live on top of each other in tiers.

In the North Island, the shallow waters around the Firth of Thames and the Waihou River were a famous flounder ground, which led to fighting between different hapū in the area.[7]

Black flounder (*Rhombosolea retiaria*) by Frank Edward Clarke, 1870. (*Te Papa, 1992-0035-2278/19*)

A SYMBOL OF ABUNDANCE

Because of their great abundance, flounder became a symbol for bountiful food and the ability of communities to sustain themselves. Artists developed a pattern known as pātikitiki, based on the iconic diamond shape of the flounder, which represented the bounty that the sea can provide, and wove it into kete, cloaks and tukutuku panels. The diamond shape of flounder was also the inspiration for children's kites, and in some traditions Te Ika a Māui (the North Island) was believed to be a flounder – a great diamond shape that stretches from Cape Reinga to Wellington and from East Cape to Taranaki. Pātiktiki patterns remain in use in contemporary artwork, and are found today in weaving, carving and in building and street pavement designs.

A kete with a pātikitiki weaving pattern. *(Auckland War Memorial Museum, Tāmaki Paenga Hira, 32202)*

FISHING FOR FLOUNDER

Flounder was immediately a favourite of British settlers in New Zealand, who prized it as one of the most delicious fish in the country. They collected the fish by wading estuaries at night armed with a bright light and sharp stick, by dragging a net across shallow bays and harbours, or even by stabbing the fish over the side of a boat as they were spotted. Flounder were a useful food supplement when supplies were low, and could be caught even with little fishing experience or access to equipment.

During the New Zealand Wars of the 1860s, British soldiers were known to use their bayonets to catch flounder to supplement their rations. Flounder were so popular with European settlers that as early as 1860 there were observations that flounder were becoming smaller and scarcer, and there was a growing awareness that they needed protection. Before fishing regulations were imposed, there was nothing to stop very tiny juvenile fish from being sold. There are reports from the 1880s that a Dunedin hotel served flounder that were less than 5 centimetres long.[8]

Kaipara Harbour was a particularly popular flounder-fishing ground; the flounder were so abundant that the bottom of the harbour was said to resemble a forest of autumn leaves, with juvenile flounder covering the sea floor. Even as late as the 1960s, fishers could spear up to 140 flounder on the incoming tide, and in the 1970s fishers recalled barely being able to set foot on the sand without disturbing flounder, and picking them up out of the water by hand.

In some areas flounder were once so abundant it was difficult to walk without stepping on them. *(Luke Colmer)*

While flounder are no longer as abundant and large as they once were, flounder fishing remains popular today. It is arguably one of the easiest forms of fishing, as the fish can be caught in knee-deep water or by 'drifting'– searching for the fish as you are carried along by the tide. The use of underwater lights and dive torches makes spotting them at night much easier, and some fishers have 'turbo-charged' their equipment by using high-powered beams hooked up to old car batteries. It is not unheard of to get a 'fish kebab', with two or more flounder on the spear at a time, as the fish will sometimes swim over one another when feeding. People often stumble across flounder unintentionally while swimming, as they refuse to move until the very last moment.

Yellow-belly flounder (*Rhombosolea leporina*) by Frank Edward Clarke, 1870. *(Te Papa, 1992-0035-2278/28)*

BIOLOGY

For such an awkward-looking fish, flounder are remarkably efficient swimmers, able to dart about the sea floor with rapid bursts of speed. They can even travel surprisingly large distances; one flounder was tracked travelling from Lake Ellesmere to Foveaux Strait – a distance of over 400 kilometres. They have incredible powers of camouflage; their skin is studded with colour pigments which they can contract to change colour to match their surroundings. Flounder caught in a net can sometimes recreate the pattern of the net across their skin and if a chess board is placed in a tank with a captive flounder, it will take on a chequerboard pattern.

Flounder are masters of camouflage, changing pigments in their skin to match their surroundings. *(Shaun Lee)*

Stingray / Whai

Guardians of the shore

TAXONOMY

The most common stingray species in New Zealand are the short-tailed stingray (*Bathytoshia brevicaudata*), the long-tailed stingray (*B. lata*) and the eagle ray (*Myliobatis tenuicaudatus*). The easiest way to tell stingrays and eagle rays apart is that stingrays have round bodies that ripple as they move while eagle rays flap their fins like a bird's wings.

Stingray glide along the sea floor like flying carpets, hunting out crabs and seashells with the electrical receptors around their jaw. These otherworldly creatures were believed to be spiritual guardians – kaitiaki – protecting the shellfish beds of harbours and estuaries. If everyone followed the correct tikanga while fishing, then the stingray would make sure there were plenty of shellfish for everyone. But if an enemy group arrived with evil intent, then these kaitiaki had the ability to capsize ships with rogue waves and send their crews to their deaths.

From all around Aotearoa come tales of the important role stingrays played as guardians and protectors. One tradition from Tāmaki Makaurau/Auckland tells of a tohunga named Hape who was refused a place in a waka on the voyage to Aotearoa because of his club foot, so he caught a ride on the back of a giant stingray instead. Once Hape arrived in Aotearoa, the stingray known as Kaiwhare took up residence in Manukau Harbour, where it continues to protect the area today.

A short-tailed stingray glides over a forest of brown kelp (*Ecklonia radiata*). (*Sarah Milicich*)

A DEADLY SPEAR

While many stingrays were regarded as tapu and protected, they were also seen as a valuable food source and were held up with sharks and kingfish as one of the best-tasting seafoods. They could be speared in shallow harbours, swept up in larger beach nets, or caught on hook and line and then dried on racks in the sun.

The most famous stingray of all was caught by the demigod Māui, using a magical hook formed from his grandmother's jawbone and using blood from his own nose as bait. The great stingray, known as Te Ika a Māui, now forms the North Island of New Zealand. Its head is at Wellington (Te Upoko o Te Ika), its fins are at Taranaki and East Cape, and its long tail stretches to Cape Reinga (Te Hiku o Te Ika), with the stinging barb forming the Coromandel Peninsula (Te Tara o Te Ika).

Māori were well acquainted with the stinging properties of the barb of cartilage at the base of the stingray's tail, which is covered in a toxic mucus that causes body tissue to wither and die. It was believed that Tangaroa gave the barb as a gift to the stingrays for their valour in battle. After a battle between humans and fish, Tangaroa let the leader of the stingrays choose anything he wanted from the spoils of war for

Short-tailed stingray (*Bathytoshia brevicaudata*) by A. R. McCulloch (1911), adapted by Lars Quickfall.

himself and his tribe. The stingray leader selected one of the humans' spears with a row of barbs and asked Tangaroa for a tail just like it.

Māori used these barbs to make throwing spears – tara whai – which could be over 2 metres long and were used when defending a pā against attack.[9] The barbs would break off and lodge in the victim, and were most likely intended to maim and slow down an enemy rather than deliver a fatal blow. The vicious hold of a stingray barb inspired the whakataukī 'He tara whai ka uru ki rote, e kore e taea te whakahokia' ('A stingray's barb, deeply thrust in, cannot be withdrawn'), which was used as a metaphor to describe an idea that has taken hold in the mind, or a grudge between people that was difficult to overcome.

ETYMOLOGY

The scientific name *Bathytoshia* is a combination of 'bathy' meaning deep and 'tosh', in honour of marine biologist James R. Tosh. The name *Myliobatis* means 'grinding flat fish', a reference to the eagle ray's wide grinding teeth, and *tenuicaudatus* means 'slender tail'. The Māori name 'whai' originated in the Pacific, where similar words such as 'fai' or 'hai' are used for stingrays. In Māori 'whai' also has another meaning – 'to chase or pursue'. The collective noun for a group of stingrays is a 'fever'.

THE STING IN THE TAIL

On rare occasions, Pākehā indulged in stingrays as food, especially the wings, which produce firm, tasty fillets with no bones. But for the most part, British settlers in New Zealand generally disliked these armed fish with their stinging tail, and there was a long-held belief that their sting contained a poison sac, like a bee's. Commercial fishermen using set nets especially considered them to be a pest, as they would get tangled up and damage their nets and were difficult to remove safely. Rather than risk encountering them again, some fishers preferred to cut off their tails before throwing them back.

A fever of stingrays, gliding through the clear blue waters of the Poor Knights Islands. *(SeacologyNZ)*

In 2006 a tragic fatal stingray attack shocked the world, and forever changed the public perception of stingrays. Wildlife filmmaker Steve Irwin approached a short-tailed stingray from behind while filming a documentary series in Australia. The stingray suddenly turned and began thrusting its barb wildly, stabbing Irwin in the chest dozens of times and fatally injuring his heart and lung. It is possible that the stingray mistook Irwin for a tiger shark, its main predator in Australia.

Only one similar stingray fatality has been recorded in New Zealand waters. In 1938, eighteen-year-old Jessie Merle Laing was wading in shallow water at Te Mata near Thames when she was struck in the chest by a stingray barb. Her fiancé Frederick Banfield flagged down a truck to take them to Thames Hospital, but sadly Jessie didn't make it in time. Such attacks are extremely rare worldwide, however, and although stings can be very painful, they are not usually life-threatening unless a vital organ is pierced. Most of the time stingrays are calm and peaceful, and content for divers to swim with them. They can be even made quite tame through feeding, and at Tatapouri near Gisborne feeding short-tailed stingrays and eagle rays has become a popular tourist attraction.

BIOLOGY

The short-tailed stingray is the largest stingray in the world and can weigh over 200 kilograms. It has large, plate-like teeth for crushing up molluscs and other prey. During the mating season, the males' teeth change shape and become sharper so they can grip on to females. Sometimes stingrays gather in huge numbers in places such as the Poor Knights Islands off the Northland coast to breed, swirling and flying together in huge flocks before dispersing.

FAVOURITE FARE

New Zealand stingrays have the unfortunate distinction of being the favourite food of the ocean's top predator – the orca. Around the world orcas are sustained by a variety of food sources, but only in New Zealand waters do they rely on stingrays as their staple diet. The whales scour the sandy bottoms of harbours, hunting stingrays with their sonar. Once they have located them, the pod works together to grab the rays then flip them onto their backs, while carefully avoiding their barbs. Stingrays have a very primitive nervous system, and once turned upside down they enter a state known as tonic immobility, where they are effectively paralysed and can't defend themselves. They are then easily devoured by the orcas, which sometimes toss the helpless rays out of the water like Frisbees. To hunt down more elusive stingrays, orcas have been observed pinning them to the ground or blowing bubbles at them to force them away from their hiding spots.

But stingrays do not give up without a fight, and have been known to injure and even kill orcas in defence. One young orca was found dead floating near the Noises Islands with a barb lodged in the back of her throat, having died either from blood loss or a reaction to the stingray toxin.[10] Sometimes stingrays will race into very shallow water and even intentionally beach themselves to avoid capture. Many of the places orcas strand themselves are in stingray hunting grounds, and it is thought they may run themselves aground during an attempt to catch stingrays.

A short-tailed stingray, seen from below. *(SeacologyNZ)*

An eagle ray (*Myliobatus tenuicaudatus*) rests among kelp. *(Shaun Lee)*

Gurnard / Kumukumu

Grunting sand-stalkers

Gurnard (*Chelidonicthys kumu*) by Antoine Germain Bevalet, 1824. *(Smithsonian Libraries)*

Gurnard are the butterflies of the ocean, with orange-red bodies and delicate wing-like fins coloured a mixture of fluorescent blues and greens, metallic hues and sky-blue spots. But their beautiful appearance is deceptive, as gurnard are efficient and deadly hunters. On their chins are modified fins that help them 'walk' along the sea floor. As each fin probes into the sand, it detects the slightest movement. When gurnard find a creature to eat, they disturb it with their fins then gobble it up as it tries to escape.[11]

THE GROAN OF DEFEAT

One of the most intriguing features of this fish is the loud grunting noise it makes. One traditional Māori explanation says the sound originated from a great battle between fish and humans. The story goes that there was once a man who abandoned his wife and ran away to a neighbouring tribe, so the wife called on Tangaroa for help. Tangaroa heeded the call and raised an army of fish to make war on those sheltering the runaway husband. The gurnard tribe led the assault on the humans, but they were quickly slaughtered. The few survivors were stained red with blood, and doomed to forever groan in anguish over their fallen comrades.

Another version of the tale has a different end to the story, in which the gurnards were victorious in battle and allowed to take anything from the battlefield as a reward. The leader of the gurnards was intrigued by the blood and sound of groans from the dying men, so he asked Tangaroa for his tribe to be coloured red and to be given the ability to make grunting noises.

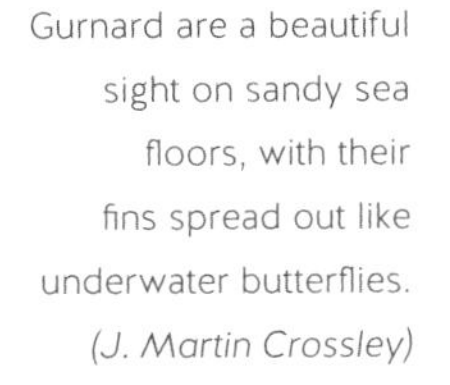

Gurnard are a beautiful sight on sandy sea floors, with their fins spread out like underwater butterflies. *(J. Martin Crossley)*

BEATING THE DRUM

Today most people notice the gurnard's grunting noises when they catch these fish, and a boat full of gurnards produces a bizarre chorus that has been compared to the sound of a pigsty. It is produced by a special type of 'drumming muscle' that beats against the swim bladder.

Scientists studying New Zealand red gurnard found they produce a wider range of vocal noises than any other fish in the Triglidae family, and grunt at all times of the day and night. They detected two distinct sounds the fish can produce – a grunt and a growl. Gurnard tend to growl when they are swimming on their own at night, whereas the grunts are more common when a group of gurnard get together, and may be used to signal their reproductive status. During the breeding season, the ocean soundscape can be filled with the noise of grunting gurnard looking for mates.

TAXONOMY

Red gurnard or kumukumu (*Chelidonichthys kumu*) is a member of the sea robin family Triglidae. This group of fish all have large pectoral fins that resemble wings and possess a 'drumming muscle' which they use to produce grunts and growls. Their relatives in New Zealand are the scaly gurnard (*Lepidotrigla brachyoptera*) and the spotted gurnard (*Pterygotrigla andertoni*).

Although red colours are associated with rangatira, gurnard were often regarded as an inferior food. *(Brian Gratwicke)*

Illustration of gurnard by Franz Doflein, 1910. *(Smithsonian Libraries)*

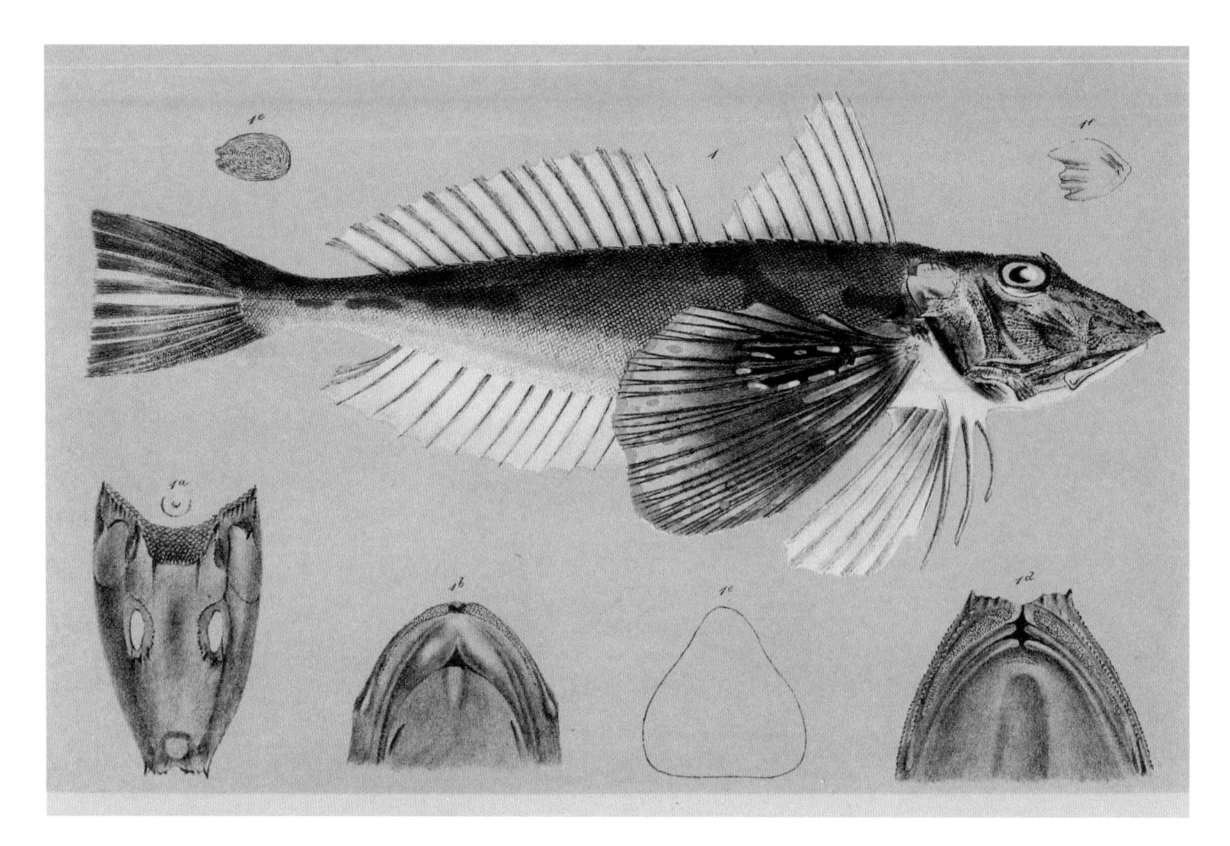

Illustration of gurnard by Arthur Bartholomew, 1885. *(Biodiversity Heritage Library)*

RELATIVE OF REPTILES

Gurnard were a useful food source for Māori in many places around the country, but in some areas the fish was considered a lowly or inferior food. One story goes that a group of fishermen spotted the rangatira Kahukura walking along a beach in Taranaki, covered head to foot in red ochre. One of the men caught a gurnard and made a joke, comparing Kahukura to the red fish. However, news of the joke made it back to Kahukura, who was gravely insulted at being compared to a lowly fish, and immediately launched a devastating attack on the joker's village in revenge.

Possibly the poor reputation of this fish arose from it being associated with reptiles, which were often regarded with loathing and fear. In one traditional whakapapa, gurnard were the children of Punga – the atua of reptiles and ugly things.[12] It was said that when Mahuika – atua of fire – was raging through the land burning the forests, gurnard called on his brother tuatara to join him in the ocean. But the tuatara scolded the gurnard, saying, 'I will stay on land

ETYMOLOGY

The name *Chelidonicthys* roughly translates as 'a fish like the swallow', and refers to the large pectoral fins of this genus, which resemble bird wings. The species name *kumu* comes from the Māori name for this species, kumukumu, a word that originally referred to groaning and grunting sounds. The English name 'gurnard' also means 'to grunt', and comes from the Latin *grunire*. Gurnard are sometimes referred to as 'carrots' due to their striking orange-red colour.

BIOLOGY

Gurnard live in soft, sandy sediments all around New Zealand, in water up to 150 metres deep. Their strong orange-red colouration is especially vibrant when they are stressed, such as after they have been hooked and pulled into a boat. Underwater they can be a blotchy red or brown colour, with a pale white belly. It is thought that their beautiful fins may be used to startle predators, attract mates and to help keep them stable as they swim.

where all will fear my ugly appearance, whereas you will be caught by humans and served in a basket of food at a feast.' His prediction came true, as the gurnard was frequently eaten at feasts, whereas tuatara were feared and very rarely eaten.

Pākehā, too, had a strong prejudice against gurnard as an eating fish for many years, perhaps related to its unusual appearance. There was very little demand for gurnard in New Zealand, and they were generally sold cheaply and bought only when snapper was unavailable. When gurnard was eaten, it was baked, boiled or fried, and the flesh canned or preserved in casks. Some recipes called for it to be soaked in white wine, or stuffed with forcemeat baked with bacon.

The lack of interest in gurnard in New Zealand meant these fish were often exported to Australia, although there was some prejudice against the fish there as well. Today, however, gurnard has undergone a complete reversal of fortune: it is now considered one of the finest eating fish in New Zealand, and is many people's favourite fish. It is a staple at fish and chip shops, and is served in high-end restaurants as well, with the fish being poached, baked, steamed, grilled, fried, cooked in soups and chowders, or served as ceviche or sashimi.

The prejudice against consuming gurnard has waned, and it is now highly valued as an eating fish. *(J. Martin Crossley)*

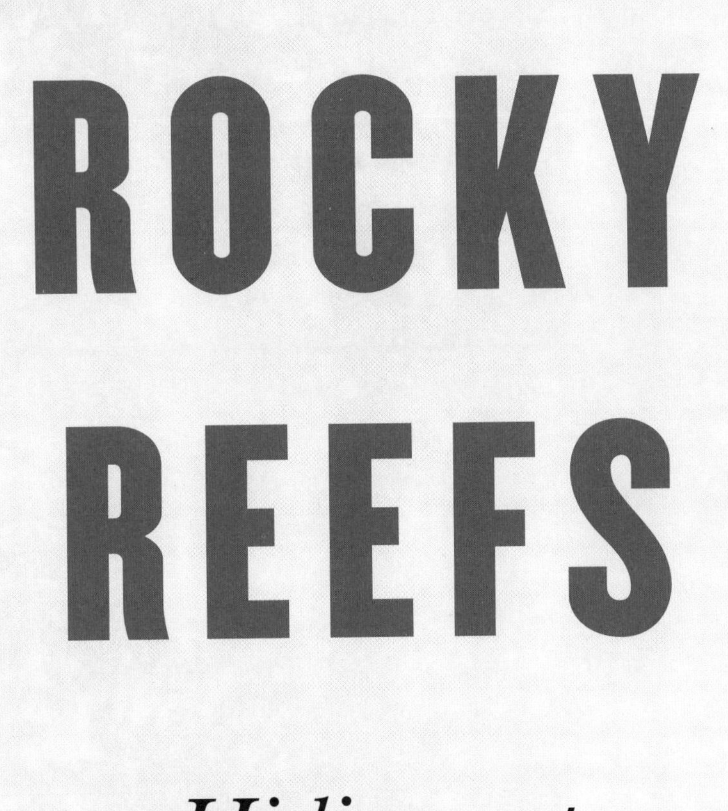

ROCKY REEFS

Hiding out in fortresses and forests

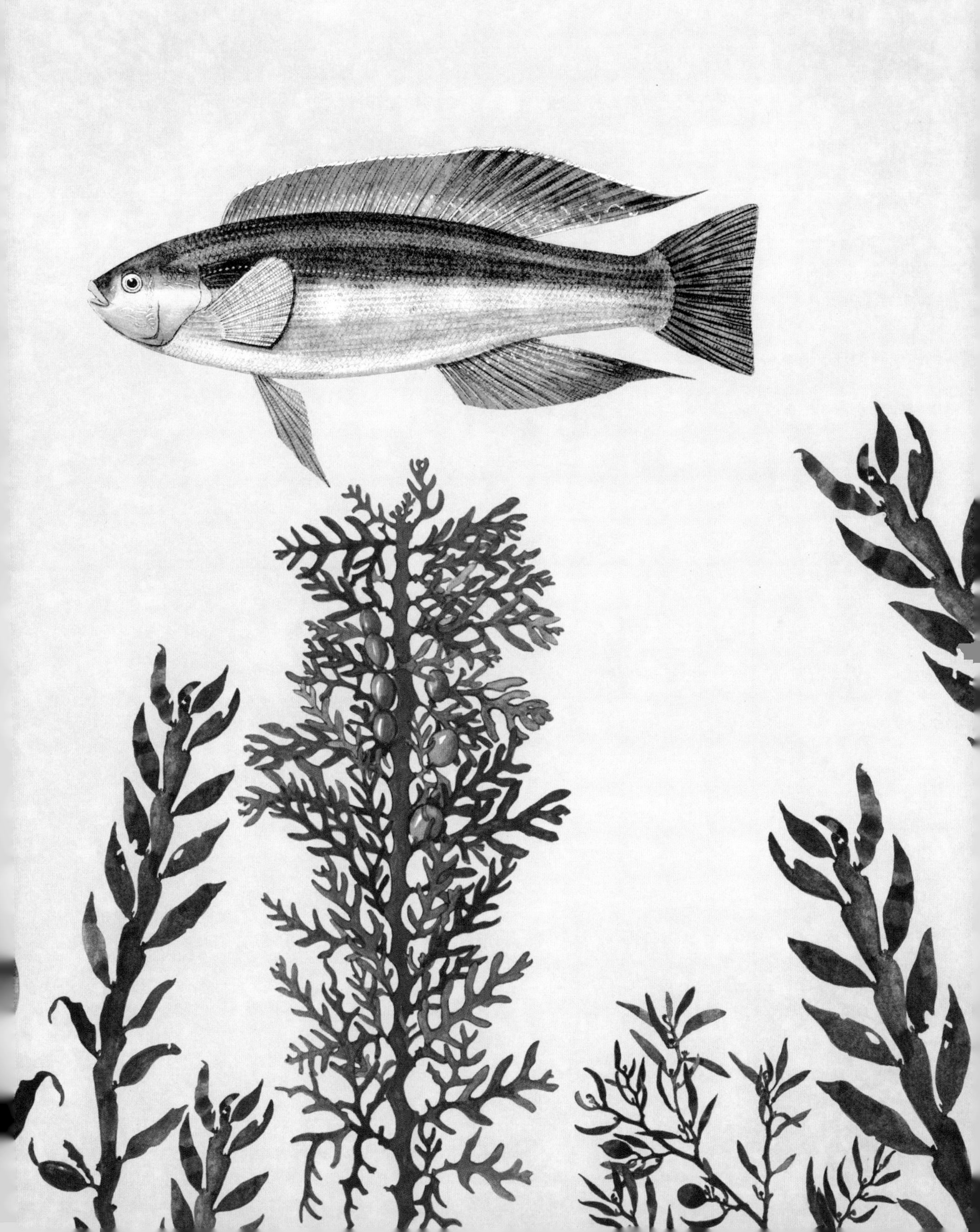

Crayfish / Kōura

Spiky sea bugs

TAXONOMY

The most common saltwater species of crayfish are the red rock lobster (*Jasus edwardsii*) and the larger packhorse lobster (*Sagmariasus verrauxi*), both of which can also be found in Australia. They belong to the spiny lobster family Palinuridae, which all have large horns and antennae but no claws.

Crayfish are like gigantic underwater insects covered in a suit of armour. Every inch of their fearsome bodies is studded with spines and bristles, and they have two long, probing antennae that can detect the slightest movement in the water around them.

In the daytime, they conceal themselves in caves, cracks and crevices. These guarded fortresses can be occupied by up to fifty crayfish, their spiky antennae interlocking so that no predator can enter without them sensing it. At dusk, when their enemies are less active, crayfish emerge to stalk the sea floor, devouring just about everything in their path. What they don't eat right away they will collect for later, scooping up shellfish and taking them back to eat in their rocky lair.

A crayfish guarding its rocky lair. *(Luke Colmer)*

CHILDREN OF THE ROCKS

In Māori tradition, crayfish were the offspring of Rakahore – the personification of rocks and stones – and were brought down from the heavens by the atua Tāwhaki and released into the ocean. Once there, crayfish multiplied and provided a highly abundant and valuable food source.

There were several methods of catching crayfish. In many places they could simply be plucked out of rock pools at low tide. But where they were out a bit deeper, they could be lured out of their rocky crevices with a bit of pāua dangled on a string, or into traps constructed from vines of supplejack and mangemange and baited with kina and shellfish. But perhaps the most impressive technique was to dive for them.

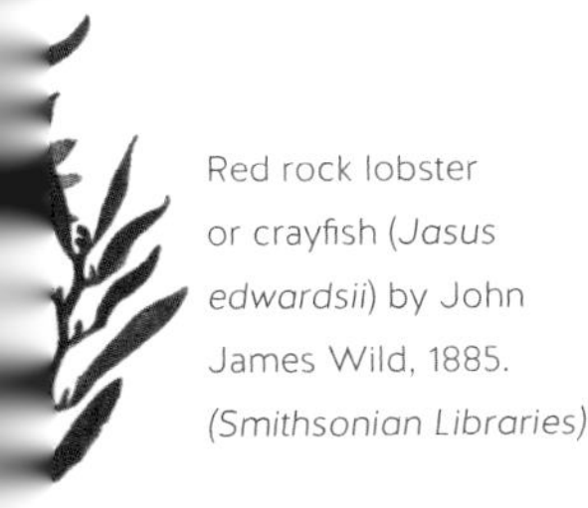

Red rock lobster or crayfish (*Jasus edwardsii*) by John James Wild, 1885. *(Smithsonian Libraries)*

The traditional method of diving for crayfish – ruku kōura – involved diving feet-first 5 metres or more and feeling for the crayfish with the feet. Then the crayfish could be manoeuvred into flax kete whose drawstrings could be pulled closed, like a catch bag. Great crayfish divers were highly admired, and women were often considered better at the pursuit than men.[1]

Once caught, crayfish were smoked, cooked in a hāngī or threaded on strings of flax and dried in the sun. Another preparation method was to tie a bundle of crayfish together and leave them in a freshwater stream, weighted down with a branch or stone. Over time the meat became soft and could be slipped out of its shell. The meat was then eaten raw, cooked in a hāngī or sealed in a bull kelp bag for later use.

To protect crayfish populations from overharvesting, if a crayfish was discovered carrying eggs the area was put under a rāhui to prevent catches during this important breeding time.

ETYMOLOGY

The meaning of the name *Jasus* is unclear, but it might refer to the Latin name *Iasus*, for an ancient town in Turkey; the name *edwardsii* was given in honour of the nineteenth-century French crustacean expert Alphonse Milne-Edwards. The name *Sagmariasus* comes from a Greek word for a packhorse, as the sides of the adult lobster bulge out as if it is carrying a load on its back. The name *verrauxi* is likely a reference to the French natural historians Jules and Édouard Verreaux.

The name kōura is used across Polynesia as a term for crayfish, shrimp and lobsters. Many places in the country are named after kōura, such as Kaikōura. Crayfish are also known as 'crays' or 'bugs'.

A crayfish pot, or taruke, made primarily from vines of supplejack and mangemange. *(Auckland Museum, 12.28062-i)*

LUXURIOUS LOBSTERS

When the first European explorers arrived in New Zealand, their favourite food to trade for with Māori was crayfish. Joseph Banks, the naturalist on board HMS *Endeavour*, wrote that 'above all the luxuries we met with, the Lobsters or Sea Craw fish must not be forgot'. He frequently traded with Māori to obtain them and believed they were 'certainly the largest and best I had ever eat'.[2] Other European explorers recorded catching so many crayfish they didn't know what to do with them all, and noted that if they left their fishing nets in the water too long their catch would be completely consumed by crayfish. Those based at remote whaling stations would catch crayfish and cure them to use as food over the winter months. Missionary James West Stack ate so many crayfish when he visited Ōnuku marae in Akaroa Harbour in the 1860s that he made himself ill. During his stay he was treated to crayfish for breakfast, lunch and dinner for two whole days. He wrote, 'At first we ate freely of them, for we all liked them. But after our third meal we began to suffer from headaches, which grew worse and worse after every additional meal.'[3] When Stack asked his Māori hosts what was happening, they explained that it was a common side-effect of eating a pure crayfish diet.

A FOOD FOR DRUNKARDS

While early explorers and settlers saw crayfish as a luxury, the next wave of settlers had a completely different view of them. Many regarded New Zealand crayfish as tasteless compared to English lobsters. Worse still, because crayfish were sold ready-boiled at markets there was a widespread belief that they were the food of drunkards, the poor and those too lazy to cook. For a long time, crayfish was one of the cheapest seafoods available and occasionally excess crayfish that couldn't be sold would have to be buried or disposed of at sea. In the north, huge hauls of packhorse crayfish (*Sagmariasus verrauxi*) were taken and while some might be kept as food, the rest were fed to chickens. Around the coasts of Ngunguru in Northland, the nets of commercial mullet fishers would become so overrun with crayfish eating their fish that they would often crush them and dispose of them – a form of pest control.

And yet, for those less prejudiced, there were still huge numbers of crayfish to be had right from the shore. Māori children were known to pick up crayfish from shallow water on their way home from school, grabbing one for each family member. In many places around the country, crayfish were so abundant that at low tide their feelers would stick out of the water. During World War II, crayfish were sent to soldiers serving in the Māori Battalion in Egypt, in tins sealed with lard. The soldiers described these deliveries as a godsend, spread the rich meat on bread and blessed the whānau that had sent it.

Crayfish were once so abundant that some people even viewed them as a pest. *(Ian Skipworth)*

RED GOLD

Overseas markets such as America began to develop a taste for New Zealand crayfish, and the growing demand led to an unprecedented fishing boom. In 1949 alone, 18 million crayfish tails were sent to the United States. Commercial fishers across the country began converting their boats and gear to target crayfish, to make the most of the 'red gold' swarming across New Zealand reefs.

But many fishing practices were unforgivably wasteful: only the tails were taken and the heads and legs tossed back in the water. With the potential to make huge amounts of money, some would get creative to avoid fishing rules. In Karitane in Otago, undersized crayfish were sold under the name 'Karitane rock lobster', the vendors claiming it was a separate, smaller species not subject to normal fishing regulations.

As crayfish numbers around the mainland began to diminish, fishers flocked to the Chatham Islands, where crayfish were still hugely abundant. On any given day, hundreds of fishing boats could be seen laying out crayfish pots. But by the 1970s the fishing boom was ending, as crayfish populations continued to decline and their slow rate of growth meant they couldn't recover quickly.

BIOLOGY

Crayfish go through incredible migrations over the course of their lives. After mating, the female protects the developing eggs underneath her tail for several months, until they are ready to hatch in spring. The larvae drift hundreds of kilometres out to sea, where they spend over a year floating in the open ocean. Sometimes they are swept up into big, swirling ocean gyres, before finding their way back to rocky reefs to settle. Adult crayfish have also been observed making huge underwater migrations, somehow navigating back to their spawning sites. In the 1970s crayfish were tracked walking from Otago to Fiordland, crossing the sea floor in long files with as many as 50 lobsters in a group, keeping together by connecting with their long antennae.

A crayfish peers out from a dark crevice. *(Daan Hoffman)*

FUNCTIONAL EXTINCTION

Today, while crayfish harvests are much better managed, many populations around the country are continuing to decline. Scientists have even observed huge drops in crayfish numbers inside fully protected marine reserves. In New Zealand's biggest marine park, the Hauraki Gulf, the crayfish population is estimated to be at less than ten per cent of historic levels and crayfish are now regarded as 'functionally extinct' – meaning there are so few left they are unable to perform their usual role in the ecosystem.

Because the numbers of crayfish and other predators are so low, kina populations in the gulf have exploded. In turn, diverse kelp ecosystems have been replaced by kina barrens devoid of life. Many scientists, conservationists, fishers and iwi have called on the government to extend current marine reserves and review crayfishing regulations to protect and restore crayfish populations.

Kina / Sea urchin

Hedgehog of the sea

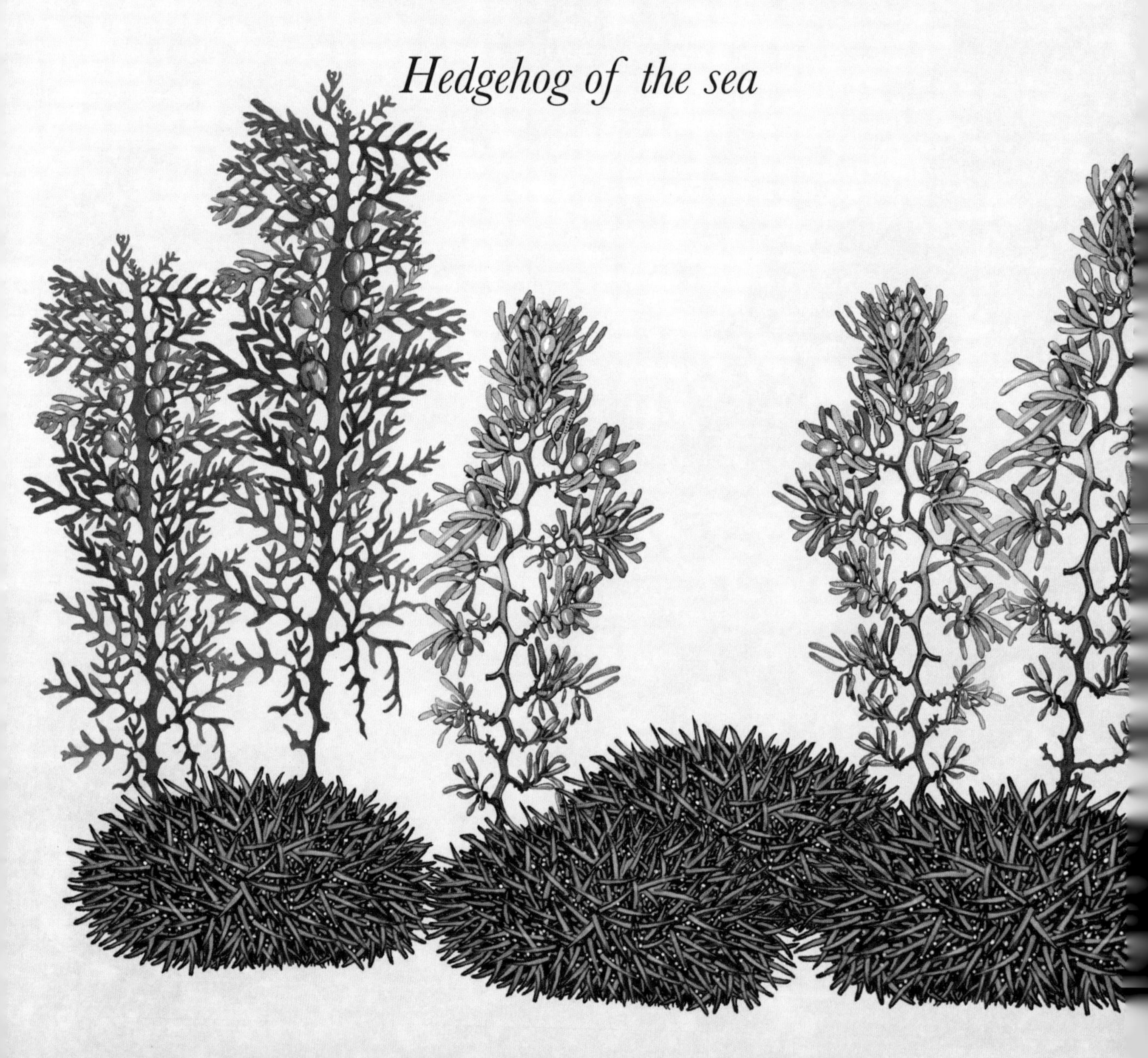

The tiny grinding mouthpiece of the kina is surrounded by sharp spines. *(SeacologyNZ)*

Kina resemble something from a nightmarish steam-punk fantasy – marching across the sea floor like mechanical engines designed for felling kelp forests. They are armoured with hundreds of spines that spin around in all directions, protecting them from attack. In between the spines are tube feet – sticky tentacles operated by hydraulic pumps – that kina use to pull themselves along. Their feeding structure is a masterpiece of engineering, with forty different components operating powerful chomping teeth. These teeth can cut through the toughest algae and are regenerated as soon as they are ground down.

THE KINA BARRENS

Most of the time, kina are content to hide in crevices, covering themselves in rocks for camouflage and catching pieces of floating kelp to feed on. But when their main predators – snapper and crayfish – are overfished in an area, their numbers explode. With no reason to hide, they unleash an all-out assault on the kelp forests and begin munching directly on kelp stalks. During this feeding frenzy, kina even climb the kelp to nibble on their blades, eventually toppling the plants to the sea floor, where more kina pile on.

In the aftermath of such destruction, huge areas of reef are stripped bare of kelp and left as marine wastelands known as 'kina barrens'. Without the food and habitat of diverse kelp forests, the ecosystem collapses as fish, invertebrates and algae are wiped out in the area. Only a small number of kina are needed to stop new kelp growing and maintain a kina barren. Fortunately, however, no-take marine reserves such as Goat Island north of Auckland show that the trend can be reversed; when large snapper and crayfish are protected, lush kelp forests can return over time.

Kina (*Evechinus chloroticus*) by Arthur Powell (1947), adapted by Lars Quickfall, with kelp (*Carpophyllum plumosum* and *Cystophora torulosa*) by Nancy Adams. (*Te Papa*)

A moving front of feeding kina can quickly strip a kelp forest bare if left unchecked. *(Ian Skipworth)*

PRIZED ROE

Kina provided Māori with a delicious and readily available food source, collected in the shallows and eaten raw, straight from the sea. While all of the innards were eaten, most prized part was the roe, which in spring, when the kina are ready to reproduce, swells to become bright orange, like slices of a mandarin. Out of season, the roe can be very bitter, so it was important to harvest at the right time of year.

The blooming of the kōwhai tree was a signal that the roe were starting to develop, although some believed they were at their sweetest when the pōhutukawa were in flower. A method for preparing the kina roe that is still used today was to place it in fresh water for several days, as it then becomes even sweeter and loses its salty ocean tang.

When kina were cooked, the shell itself made an excellent cooking vessel, and it was filled with kina roe and placed on the burning embers of the fire. When left on the beach to dry in the sun, these shells lose all their spines and become a pretty pale-green colour. They make beautiful ornaments, and were used as containers for storing perfumes and fragrant mosses.

TAXONOMY

There are around seventy species of sea urchin in New Zealand waters, with kina (*Evechinus chloroticus*) being the most common and well known. They belong to the phylum Echinodermata, which includes starfish and sea biscuits. If you compare kina and starfish, they have the same basic body plan. Starfish have five or more arms covered in tentacles called tube feet, which help them move. In kina, the tube feet sit between their spines and the arms have fused together to form their skeleton.

The beautiful shell or 'test' is what remains of the skeleton once the spines have fallen off. *(Daan Hoffman)*

Pākehā also admired the pale-green shells found washed up on beaches, and they have become icons of coastal New Zealand, used as ornaments, soap containers and jewellery boxes. Today they remain a popular symbol throughout New Zealand and are a common feature in artwork and sculptures. They were not often eaten by Europeans in the early days of settlement, but were sometimes relied upon by those living in remote areas who were forced to live off the land. Charles Heaphy, on his travels along the west coast of the South Island in the 1840s, ate them to supplement his diet; he described them as palatable enough, but noted they 'would be much improved by vinegar and condiments'.[4]

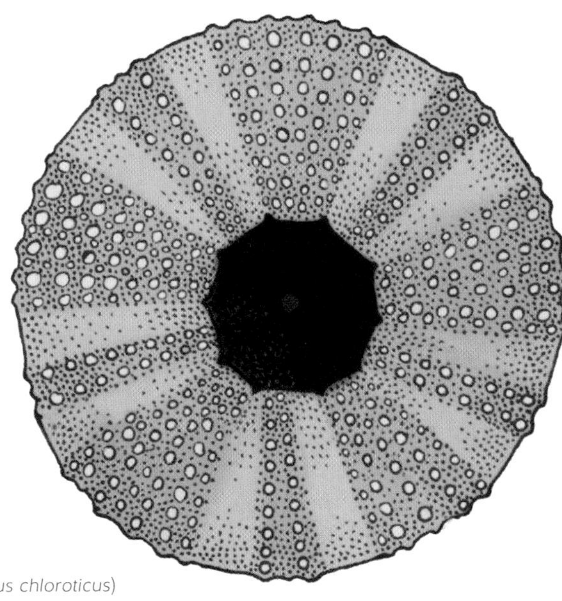

Kina (*Evechinus chloroticus*) test by Arthur Powell (1947), adapted by Lars Quickfall.

As time has gone on, more New Zealanders have acquired a taste for kina roe, some even going as far as to call it 'the caviar of New Zealand'. It has been described as the whole sea in a mouthful, and can taste salty, earthy, sweet and savoury all at once. It's now a staple of modern New Zealand cuisine, and there are recipes for kina omelettes, kina fritters, raw kina salad, kina ice cream, and even a classic Kiwi kina dip made with sour cream, cream cheese and citrus.

With more interest in eating them, kina diving has become a popular sport. Tools for harvesting kina off rocks are often improvised out of screwdrivers, garden hoes and barbecue scrapers. Fishers value kina as burley and bait, crushing up the shells to lure fish to an area. Kina are also a critical tool in the arsenal of spear fishers, who use piles of crushed kina as bait and lie in ambush in kelp, waiting for fish to swim past.

ETYMOLOGY

The scientific name *Evechinus* means 'very spiny' and *chloroticus* means 'pale green', a reference to the dried skeleton. The family name 'Echinoderm' literally means 'hedgehog skin', from *echino* (hedgehog) and *derm* (skin). The Māori word 'kina' is also a general word for sea urchins and their relatives; for example, the sea biscuit (*Fellaster zelandiae*) is known as 'kina papa' (flat kina). The word 'kina' is used to describe sea urchins in Polynesia as well, such as the Rarotongan kina (*Echinometra mathaei*).

The sharp spines of the kina can sometimes become lodged in divers' hands, leading to infection. *(Ian Skipworth)*

SPIKY GOLD

With the growing demand for kina roe, kina fishing has become a big industry, and kina divers are able to make several thousand dollars a day. Fishing regulations prevent the use of scuba equipment, so commercial kina divers around the cold waters of Southland and Stewart Island regularly make dives over 20 metres deep on a single breath, risking blackout in pursuit of big hauls.[5]

BIOLOGY

Kina munching noises dominate the New Zealand ocean soundscape, getting louder at dawn and dusk and forming part of an oceanic chorus. The sound is much louder than you would expect given the creature's small size, their shells acting like a Helmholtz resonator, amplifying the sound in the same way that a small whistle can make a loud piercing noise. The larvae of different creatures floating out at sea have been shown to use the sound of kina munching to navigate and guide them to a suitable reef habitat to settle on.

After the roe is extracted, the shells, spines and offal are generally discarded. Recently, however, scientists have begun to investigate whether kina waste could be used in medical treatments. Kina shells are known to have bioactive properties that might help treat diabetes, heart disease and Alzheimer's disease, and extracts from kina have been shown to have antimicrobial and anti-inflammatory effects that may exceed those of current drugs on the market.

Octopus / Wheke

The devil fish

Octopus illustration by Adolphe Millot (1910), adapted by Lars Quickfall.

Octopuses are surely the closest thing to alien life on Planet Earth. They have eight tentacle arms that are an extension of their brains, and they can change their texture and colour in an instant, contort themselves into the tiniest spaces and squirt a trail of ink to avoid capture.

Octopuses are masters of camouflage. *(SeacologyNZ)*

These unusual creatures occupied a privileged position in Te Ao Māori, where an octopus was said to be responsible for the discovery of Aotearoa. Te Wheke o Muturangi was the giant pet octopus of a powerful tohunga in the spiritual homeland of Hawaiki, which would constantly steal fish from Kupe and his people. So Kupe decided to capture it, pursuing the octopus across the oceans, all the way to Aotearoa. Eventually he chased the octopus to Whekenui Bay in the Marlborough Sounds, and fought a great battle with it, slicing off its arms with his adze and killing it. Today the long, swirling channels and drowned valleys of the sounds resemble the sprawling arms of the slain octopus.[6]

TAXONOMY

There are around 16 species of benthic (bottom-dwelling) octopuses in New Zealand waters. The most frequently seen are the common New Zealand octopus (*Pinnoctopus cordiformis*), the Māori octopus (*Macroctopus maorum*) and the gloomy octopus (*Octopus tetricus*).

The distinctive white eyes and orange-coloured arms of the gloomy octopus, *Octopus tetricus*. *(Daan Hoffman)*

Octopus was considered a special food that was generally reserved as a dish for rangatira.[7] Typically, the arms were cut off and boiled or roasted on the embers of a fire. Octopuses were challenging and exciting creatures to catch, and were most often collected by using the fisher's hand or leg as bait, placing it inside an octopus's den and waiting until the creature reached out to grab it. Then the octopus could be seized and hauled onto dry land. Sometimes the fisher would even sit in the water with their back to the octopus's den and wait till it climbed aboard, then carry it piggyback-style back to the shore. To kill an octopus, fishers would twist and pull out the animal's beak or deliver a short, sharp bite between the eyes.

This 1801 illustration of the kraken, by Pierre Dénys de Montfort, was based on descriptions provided by French sailors who claimed to have been attacked by the creature off the coast of Angola.

Despite the prestige of octopuses as a food source, Māori did not admire the way that the animals could be easily dispatched when caught. The Māori word for octopus, wheke, has come to be synonymous with surrender and lack of struggle, and warriors were told 'don't die like a wheke'. However, octopuses can be tenacious fighters at times.

In the early twentieth century, as a young girl, Paea Henderson used to catch octopuses regularly around Te Araroa on the East Cape, feeling for them with her feet underneath rocks. While trying to remove an octopus from its den, it grabbed her around the neck and wouldn't release its grip. Worried about the incoming tide, she removed a flax belt tied around her waist, dived under the water and strangled the octopus with the belt, before carrying it back home.

A FATE WORSE THAN DEATH?

The first European settlers in New Zealand despised octopuses. They called them 'devil fish' and believed they could suck blood through their tentacles. This hatred may have stemmed from a long association of the octopus with the kraken – a mythical beast that was said to be able to capsize ships with its long, tentacled arms. But one publication in particular seems to have played a role in turning people against octopuses. In Victor Hugo's 1866 novel *Toilers of the sea*, a fisherman does battle against a giant octopus that attempts to suck his blood. The book portrays an octopus attack as a fate worse than death, and muses on whether the octopus provides evidence for the existence of Satan: 'The tiger can only devour you; the devil-fish … sucks your life-blood away. He draws you to him, and into himself; while bound down, glued to the ground, powerless, you feel yourself gradually emptied into this horrible pouch, which is the monster itself.'[8]

ETYMOLOGY

The word 'octopus' comes from the Greek *oktōpous*, meaning 'eight foot'. Octopuses belong to the invertebrate phylum Cephalopods, which means 'head foot' and refers to the way the arms are an extension of the head.

The words 'wheke' and 'feke' are used widely across the Pacific for octopus and squid. Wheke has multiple meanings in Te Reo Māori – it can mean 'to be outraged and angry', and it can mean 'to give up without a fight'. Octopuses, octopi and octopodes are all accepted plural forms of octopus.

The discarded remains of prey often indicate the entrance to an octopus den. *(Shaun Lee)*

British settlers brought these ideas to New Zealand, where many people seemed to harbour an extreme fear of encountering an octopus at the beach. Breathless newspaper coverage in the mid-nineteenth and early twentieth centuries described octopus 'attacks', and a bather being touched by an octopus tentacle was enough to make headlines. Reporters frequently described octopuses in exaggerated terms, calling them 'slimy, evil-looking creatures' or describing them as having 'eyes gleaming with unquenchable hate'.[9] The strong public fear of octopuses led many people to try to kill them any time they were spotted. When a giant octopus made its way into the Te Aro Baths on the Wellington waterfront, one man attacked it with a pitchfork, while the caretaker ran and got an axe and started hacking off its tentacles.[10]

It certainly didn't help their reputation that octopuses are also a major predator of crayfish and can squeeze into crayfish pots, wrapping their tentacles around the cornered crayfish and sucking out their insides. They are such efficient predators that they can have a noticeable economic impact on crayfish fisheries; even if they don't kill the crayfish outright, they can tear away limbs, leaving them unsellable.

Fear of octopuses was once so great in New Zealand that octopuses would be attacked when they were spotted. *(Ian Skipworth)*

OCTOPUS RENAISSANCE

But despite the hatred many Pākehā had for octopuses, they also displayed an enduring fascination with them. In the 1930s, David Graham, the manager of the Portobello Aquarium in Otago, wrote, 'Almost every one hates and abhors this animal, yet most of the visitors will linger longer at this tank than at any others, coming back to it again and again.'[11] In more recent times there has been something of an octopus renaissance, with a huge amount of public interest in these creatures. The 2020 documentary *My Octopus Teacher*, about one man's relationship with an octopus, even won best documentary feature at the 2021 Oscars.

Although they appear so alien to us, octopuses share an intelligent curiosity about the world. *(Daan Hoffman)*

BIOLOGY

Octopuses are found on the sea floor and rocky reefs, and occasionally in tidal rock pools. Unlike humans, whose bodies primarily burn and store carbohydrates and fat for energy, the octopuses' metabolism is based around protein, and they need to eat constantly to stave off starvation. You can tell what an octopus likes to eat from the scattered remains of prey around its den.

Despite being among the most effective predators on the reef, octopuses display perhaps the most moving example of motherly love in the animal kingdom. Once her eggs hatch, the mother delicately blows water over them to keep them oxygenated, never leaving her eggs while slowly starving to death herself to give her offspring the best chance of life.

Part of this re-evaluation of octopuses has been due to the realisation that they are incredibly intelligent animals. Scientists estimate that octopuses have the intelligence of at least a small child or a dog, and they are curious, highly proficient problem-solvers that can use tools effectively. Octopuses have a concept of 'self' and can recognise themselves in a mirror or a video, and identify individual humans. An octopus at Portobello Aquarium took a dislike to a particular member of staff and would jet water at them each time they came close.

Their incredible intelligence and resourcefulness make octopuses difficult to keep in an aquarium, however, as they are tremendously good escape artists. An octopus named Inky escaped Napier Aquarium by squeezing out of a gap in his tank and making his way down a narrow drain to the ocean. Harry the octopus made his way out of the tanks at Portobello and was halfway up the steps at the nearby Otago Marine Laboratory before being spotted, while Sid the octopus made so many attempts to escape the same aquarium that the staff took pity on him and released him. Another octopus at Portobello even learned to squirt water at the lighting system, blowing the lights up every time the staff turned them on, until the keepers decided to take her back to the ocean.

Pāua

Meaty muscle of the sea

Yellowfoot pāua (*Haliotis australis*) by H. Pilsbury (1890); blackfoot pāua (*H. iris*) shell by Arthur Powell (1947), adapted by Lars Quickfall; and brown seaweeds *Lessonia variegata* and *L. adamsiae* by Nancy Adams.

Pāua often become camouflaged on the reef as their drab-coloured outer shell is overgrown with bits of algae. But these dull shells house an unusual and fascinating creature, one of our mostly highly desired seafoods, and yield some of the most beautiful jewellery in the country.

Pāua were an important resource for Māori – thick, meaty chunks of protein that could be harvested from shallow reefs and rock pools around New Zealand. Diving for pāua was a sport enjoyed by both men and women, with divers competing to outdo one another for the best haul. They had to be quick to pry the pāua off with a māripi – a flat tool made of wood or bone – for as soon as they sense danger, pāua will grip on to the rock like a vice and can become nearly impossible to dislodge.

For these stubborn pāua, one trick was to place a seven-armed starfish (*Astrostole scabra*) on top of them. The starfish are major pāua predators, and if they were put on the shell the pāua would rapidly twist back and forth in an attempt to escape. As they tried to shake off the starfish, they would loosen their grip on the rock, and the harvesters could simply pluck them off.

There were many different ways of preparing pāua. They could be eaten raw, smoked or cooked in a hāngī. They were sometimes buried until the flesh had softened into a consistency like cheese, then dug up and cut into slices or left in fresh water for several days before being eaten. To preserve them, Māori would thread them onto a flax rope and hang them out to dry in the sun, or place them in a bull kelp bag, covered in bird or seal fat. These preserved pāua were a useful food source over winter, or were used to trade with iwi living in the interior. Pāua were an important part of the rations for war parties, who could chew them as they marched. It was a sustaining food that was believed to produce saliva, reducing the need to drink water when on the move.

A drab, algae-encrusted shell of a whitefoot pāua (*Haliotis virginea*). (*Ian Skipworth*)

Generally, all parts of the pāua were eaten, including the frilled edge and the roe, although this could lead to some undesirable side-effects. This was dramatically demonstrated in one story about the ancestor Kahungunu. He was in love with Rongomaiwahine, but she was already married to the chief, Tamatakutai. Kahungunu decided to impress Rongomaiwahine with his pāua-diving ability and collected so many pāua that he filled his baskets and had to start sticking pāua to his chest.

That night everyone feasted on pāua, but Kahungunu decided to stir up trouble in the chief's marriage, so he asked only for the roe, as it was known to make you gassy. As they slept in the marae, Kahungunu broke wind and the awful stench filled the room. Tamatakutai woke up and quarrelled with his wife about the smell, angrily accusing her of having not cleaned the meeting house properly. The argument became so fiery that Rongomaiwahine eventually left him for Kahungunu.

TAXONOMY

There are three species of pāua in New Zealand: the blackfoot pāua (*Haliotis iris*) is the most common and the only one endemic to Aotearoa. Yellowfoot pāua (*H. australis*) and whitefoot or virgin pāua (*H. virginea*) are also native, and all belong to the abalone family, Haliotidae.

AN IRRESISTIBLE LURE

The interior of the pāua shell is an iridescent rainbow of blues, greens, purples and reds and has become one of the most recognisable icons of Māori culture. Pieces of pāua shell were used to make earrings and were hung on skirts and cloaks. The shells had a number of practical uses as well: if the breathing holes were stopped up with flax, they could be used as a container for oils, paints and tattoo ink. Their flashy colours made them an excellent material for trolling lures, designed to catch the attention of fish without the use of bait. They were especially important for carvers, who used pāua as inlay to add beautiful streaks of colour to wooden ornaments.

But perhaps their most important use was as the eyes of carved human and animal figures. They were said to give life to the carving and were thought to resemble the eyes of the ruru (morepork), keeping a watchful lookout at night. Shells with red colours were particularly sought after for carvings, as red was considered a chiefly colour and gave a fierce expression to the carvings, like a blaze of fire.

ETYMOLOGY

The name *Haliotis* means 'sea ear' and *iris* means 'rainbow'. The name 'pāua' is used in the Pacific to refer to a number of different shellfish, including mussels and clams. It also appears in placenames around the country, such as the Coromandel town of Pauanui (meaning 'big pāua' or 'many pāua'). A number of rivers have the name Waipāua, such as the Taieri River in Otago, and they were likely used as places to store pāua before they were eaten. British settlers often referred to pāua as 'mutton fish' – an apparent reference to the taste of the meat, which is something like a cross between fish and lamb. Overseas they are also known as 'abalone', 'ear shells' or 'ormers'.

The remarkable interior colours of the pāua shell are the result of genetics and the types of algae the pāua feeds on during its life.

SHINING SHELLS

Early European explorers valued pāua as a sustaining food, which could be collected easily and dried as rations. However, many later British settlers came to regard pāua meat as repulsive and wondered how anyone could bring themselves to eat it.

But although they were unimpressed by the taste of pāua, Pākehā were entranced by the colourful shells from their earliest days in Aotearoa. After James Cook returned from his first voyage to the Pacific, there was a huge demand from European shell collectors for unique New Zealand shells. Pāua was particularly prized, and on Cook's second voyage the crew were commissioned to bring them back. At the end of Cook's third voyage there were bidding wars for pāua and other shells when the ships arrived back in England, and they were traded back and forth on the dock.

British jewellers especially admired pāua shells, and in the 1800s tonnes of shells were sent to England to be made into buttons and other ornaments, with one shipment from Kaikōura exporting an estimated 150,000 pāua shells. The shells were generally collected on the shoreline already cleaned and dried by the elements, or the live shellfish were harvested at low tide with a spade and the meat discarded.

The easiest way to collect pāua shells is to find them washed up on beaches, often cleaned by the elements. *(iStock)*

After World War II, returned servicemen were given jobs processing the shells for the jewellery trade. The results were then sent overseas in an endless array of pāua-themed New Zealand souvenirs, such as earrings, necklaces, cufflinks, lamp stands and good-luck charms. While the trade put New Zealand pāua on the map, it was devastatingly wasteful, with huge quantities of pāua harvested just for their shells, and truckloads of meat dumped in the sea.

Today, pāua shells, usually collected from the shoreline by beachcombers, are so omnipresent in New Zealand culture that they have come to be seen as kitschy. They are used to ornament people's homes, hung on fences, used as ashtrays and soap dishes, and inlaid in kitchen benchtops and toilet seats.

A PRESTIGE MEAT

Pāua eventually found its way into the diets of many Pākehā. It was eaten in soups and chowders, minced and made into patties and fritters, stewed in curries or fried quickly on both sides like a steak. Pāua meat is essentially just a big muscle, so it can become quite tough, especially with overcooking, and people have developed many

BIOLOGY

Pāua are found on shallow coastal reefs around New Zealand. The colour of the inside of each pāua shell is determined by a mix of genetics and the type of seaweed they eat. While they can graze on seaweed directly, pāua prefer to graze on pieces of kelp that have been torn off the rocks and are floating in the water column. They will snatch these as they drift past, loosening their grip on the rock and grabbing them with their muscular foot. The pink 'paint' often seen covering their shells is a type of coralline algae; it releases chemicals that cause pāua larvae to leave the water column and settle on rocks, and helps promote their early development.

methods of tenderising it, such as whacking the meat with a rock, slamming it against a fence post inside a sock, or even placing it between two planks of wood and driving over them.

In the 1970s, overseas buyers got a taste for our pāua as well, prompting a fishing boom. Huge numbers of divers got into the game, and by the 1980s thousands of tonnes were being taken, before stocks declined dramatically. Today pāua is served in many high-end restaurants around the country, and a kilogram of frozen pāua can easily fetch over $100. Overseas demand is still strong, with hundreds of dollars being paid for half a dozen shellfish. In Asia pāua is regarded as a food of the elite, and at the Hong Kong restaurant Forum 'abalone king' Yeung Koon-yat serves braised New Zealand pāua for thousands of dollars a dish.

Live pāua with muscular foot, by H. Pilsbury (1890), adapted by Lars Quickfall.

Leatherjacket / Kōkiri

Gardeners of the reef

Female and male leatherjacket (*Meuschenia scaber*) by Edgar Waite (1911), adapted by Lars Quickfall.

Most fish are sleek and streamlined, to swim through the water as quickly as possible, but leatherjackets have a different strategy. Instead of moving fast, they put their resources into defending themselves from attack: they are armoured with thick, leathery skin and have a sharp spike on the top of their head, like that of a unicorn. The spine is thrust up whenever the fish is alarmed, and locked into place by a second 'trigger' spine.

Leatherjackets could never hope to chase down fast-moving prey, so instead they choose to eat prey that doesn't move, such as sponges, sea squirts and barnacles. These are foods that no other fish in New Zealand eats, as they are full of spines and toxins. But leatherjackets make short work of them, chewing off morsels to eat with their strong, chisel-like teeth and powerful jaws.

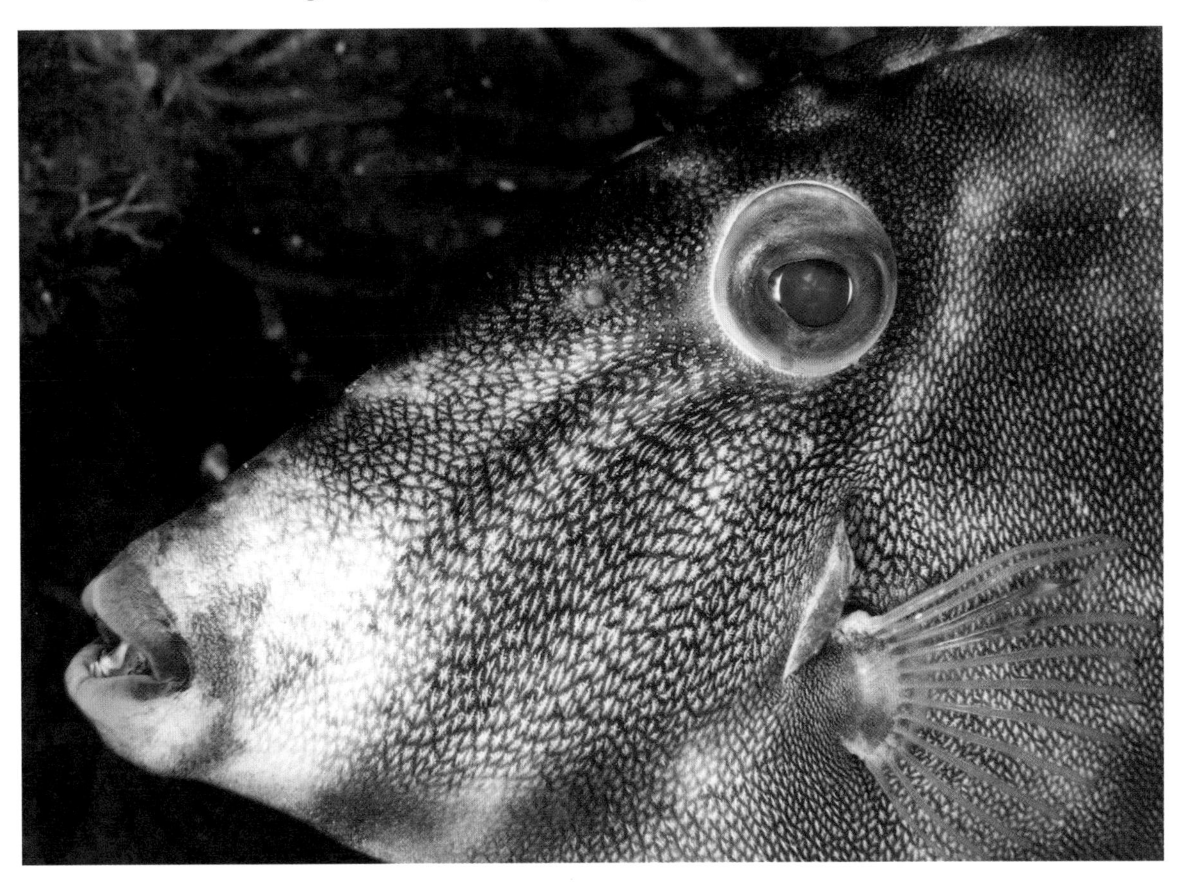

The leatherjacket's jaws are tiny, but its teeth are strong and sharp. *(Ian Skipworth)*

Their tiny teeth can have a huge impact on the reef ecosystem. Leatherjackets act like underwater gardeners, slowly pruning the surface of the reef, bite by bite. Scientists estimate that where they are abundant, leatherjackets can modify up to twenty per cent of the rock face of a reef in a year. In reefs without leatherjackets, large sponges and sea squirts dominate, but when leatherjackets are present coralline algae and super-fast-growing sponges are the main species. These sponges have developed a rapid growth rate to account for the onslaught of leatherjacket bites, and grow up to 20 times faster when being grazed.

WARRIOR FISH

With their strong defences and powerful jaws, leatherjackets are among New Zealand's boldest fish. They show almost no fear of people, and will happily eat out of a diver's hand or nibble on the end of a spear gun. Male leatherjackets are fiercely territorial and are one of the few fish that will actively bite a diver, even targeting exposed bits of skin.

Māori recognised the bold behaviour of this fish, and believed that leatherjackets were fierce warriors. In one tradition, Ruatepukepuke was said to have burned Tangaroa's house at the bottom of the sea in revenge for the death of his son, attacking fish as they fled the burning building. While most fish were badly injured and mutilated as they fled, the leatherjacket fought back. He charged out of the burning house with his throwing spear on his head and shielded by his heavy cloak of leathery skin, and got away unharmed.

TAXONOMY

The rough leatherjacket or kōkiri (*Meuschenia scaber*) is the main leatherjacket species in New Zealand waters. It belongs to the order Tetraodontiformes, which includes porcupinefish, pufferfish and sunfish. Members of this group are unable to flex their bodies from side to side and can only glide slowly using their fins, so they have developed a range of defence strategies to compensate, including rough skin, sharp spines and the ability to inflate their bodies like balloons.

Leatherjackets will bravely approach any fish or diver that enters their territory. *(Luke Colmer)*

The colours and patterns of leatherjacket skin vary so much it was once believed there were several different species. *(Ian Skipworth)*

ETYMOLOGY

The name *Meuschenia* comes from Frederick Christian Meuschen, an eighteenth-century Dutch shell collector. The name *scaber* means 'scabby' or 'rough', and refers to the texture of the leatherjacket's skin. The Māori name 'kōkiri' is used in other places across the Pacific for triggerfish such as the Picasso triggerfish (*Rhinecanthus aculeatus*). The name also carries other meanings in Te Reo Māori, such as 'to throw or thrust a spear', which is probably linked to the way the spine on the fish's head springs forward. The English names 'leatherjacket' and 'filefish' refer to its rough skin, and 'triggerfish' is a reference to the mechanism that locks its dorsal spike in place.

REBRANDING

Leatherjacket was a particularly useful food fish for Māori and was eaten from the earliest days of settlement in Aotearoa. In the north of the country, evidence from archaeological middens suggests leatherjacket may have been one of the most frequently consumed fish after snapper and mackerel. They could be caught using a tōrehe net designed for catching reef fish, and the flesh was valued highly, especially the large liver, which was considered a delicacy.

Leatherjackets were less well liked by Pākehā and developed a terrible reputation among fishermen, as their tiny mouths enable them to nibble off bait and chomp through fishing lines without getting caught. As soon as leatherjackets begin to bite, fishermen have been known to pull up anchor and try somewhere else.

If the fish was caught, it was hardly ever kept, as its strange appearance put many people off eating it.

Fishmongers sold them headed and skinned, after peeling off the leathery skin. To encourage people to buy leatherjacket, they would often sell it under the name of more popular fish like flounder. Eventually, to get away from the stigma of the name, fish-sellers agreed on a new moniker to focus attention on the delicious taste: 'cream fish'. The name has been a marketing success story, and today leatherjacket sells quite well as cream fish in markets around the country.

Even though fishers did not often eat leatherjacket, they did find uses for this fish. The thick, leathery skin makes a great substitute for sandpaper – the larger the fish, the coarser the grade – which was handy for those living in remote areas, such as lighthouse keepers, who sometimes used leatherjacket skins to sand tools and boats. The rough skin can also be used for lighting matches, and even made into shagreen, similar to sharkskin, for making decorative knife sheaths.

A juvenile leatherjacket taking refuge in kelp. *(Luke Colmer)*

A LEATHERJACKET LOVE STORY

Monogamy is pretty uncommon in the fish world – sperm and eggs are usually just broadcast into the water column. But leatherjackets seem to be a rare exception, and these bizarre-looking fish pair up to spawn.

When courting, the males show off their spine to the females, raising it up and down and nuzzling their prospective mate's leathery skin. Sometimes they use their spine to chase away other interested males. Once they are paired up, the female builds a little algae nest in which to lay her eggs, then the male fertilises them. However, leatherjacket mates might not always be completely faithful: scientists suspect the females prefer not to put all their eggs in one basket and may make nests in several males' territories.

BIOLOGY

Leatherjackets are found in shallow, rocky reefs all around New Zealand. They grow incredibly fast and are mature by one or two years old. New Zealand leatherjackets live longer than any other species in their family, with the oldest so far found to be nearly twenty years old.

When leatherjackets reproduce, their larvae settle on the tall stalks and blades of brown kelp (*Ecklonia radiata*) and grow into yellow-coloured juveniles that are camouflaged from predators. Males tend to die at a much greater rate than females, probably because they are so aggressive in defending their territories.

The rough skin of the leatherjacket has been used as a substitute for sandpaper. *(Luke Colmer)*

Parore

The black snapper

TAXONOMY

Parore (*Girella tricuspidata*) are found in northern New Zealand, and are especially abundant on the northeast coast. They are a member of the nibbler family Girellidae, which includes the bluefish (*G. cyanea*) and the caramel drummer (*G. fimbriata*).

Parore cruise around rocky reefs, grazing on seaweed like herds of wild deer in a forest. When one is spooked, the whole group takes off in a mad dash. At night, the herds break up and find sheltered places in the seaweed to sleep, their silvery grey bodies becoming a dark brown colour to help them blend in. While they are napping in the kelp, they often recruit the services of 'cleaner fish', such as young trevally and Sandager's wrasse, to nibble parasites from their skin.

Parore were a useful fish for Māori in the north of the country, where they were abundant. They could be caught using special reef nets, set nets or large seine nets drawn over sandy harbour bottoms and seagrass meadows. While parore didn't occupy as important a role in the north as other species such as snapper, they still held an important place in Māori culture and tradition. In one tale, parore fought in the great war between fish and humans, and led a band of fish into battle. Gurnard charged ahead and was slain, and parore was splattered in the blood of his fallen comrade. The blood dried and turned black, giving parore the dark stripes and colours they still wear today.

For a time, parore were loathed as spiky, nuisance fish that had no value as food. *(Erik Schogel)*

A BAD REPUTATION

Among Pākehā, however, parore developed one of the worst reputations of any native fish. There was a widespread belief that the meat tasted disgusting and had no value as a food, and a myth even developed that these fish fed on human excrement at sewerage pipes. Parore were so poorly thought of that in 1912, when several men were taken to court for dropping dynamite in Auckland Harbour and killing a school of parore, their defence was that the fish 'were practically useless as food, and nothing of value had been destroyed'.[12] In the nineteenth century, mullet and flounder fishers particularly despised the fish, as they could completely ruin a day's fishing if they were caught by mistake. When this happened, parore needed to be picked out of the nets one by one, and their spines tore apart nets and cut hands in the process. This has still been an issue in more recent times. In the 1980s, a school of around 600 parore hit a set net in the Kaiwaka River in Northland; it took a whole day to remove them and a tractor was needed to haul the net out of the river.[13] Even today, in some circles parore are still held in such contempt that fishers will refuse to let them on their boats.

Despite their poor reputation, parore can be delicious, with a firm white meat that has been compared to snapper. In fact, fish and chip shops, restaurants and

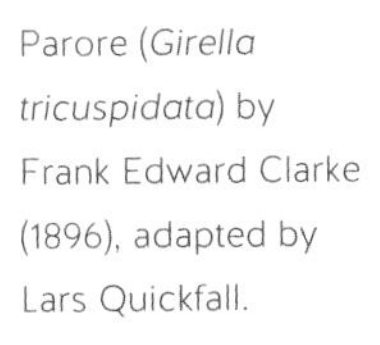

Parore (*Girella tricuspidata*) by Frank Edward Clarke (1896), adapted by Lars Quickfall.

supermarkets regularly sell large quantities of parore under the name 'black snapper'. The meat makes excellent soups and curries, and is great crumbed, fried, baked or in a raw-fish salad.

The widely varying taste of parore seems to be a result of where they live and what they eat. Parore taken from open, rocky reefs generally taste much cleaner and fresher, whereas parore taken from rivers and harbours have a muddy, weedy taste that can be extremely unpleasant and is enough to put people off for life.

Parore are often found sheltering in the seagrass meadows of shallow harbours and estuaries. *(Erik Schogel)*

THE LUDERICK

Parore also occur in Australia, where they are known as luderick or blackfish, and there could be no stronger contrast between how the fish is viewed there and in New Zealand. There is an almost cult-like devotion to parore in Australia and some fishers target nothing else. They go out in any weather, using custom-made flexible rods and purchasing packs of green seaweed as bait. At times parore has been among the most commercially important fish in eastern Australia, with strict regulations on the size and number that can be taken.

Historically, parore were seen as an everyman's fish; while trout fishing was reserved for the upper classes, parore could be fished by anyone. For many families struggling during the Great Depression of the 1930s, parore fishing provided a source of food and a welcome escape from the daily grind. One writer in the *Sydney Morning Herald* in 1905 showed the level of devotion parore could inspire in Australia, writing: 'The taking of this fish requires more skill and, in the cold days of winter more endurance, than is necessary for any other kind of fishing … To stand with the water of mid-July up to one's knees, and splashing upon one's chest, hour after hour, requires a certain kind of pluck which is not found in all men.'[14]

ETYMOLOGY

The name *Girella* comes from a French word, *girelle*, used for small, wrasse-like fishes; *tricuspidata* refers to the fish's 'tricuspid' teeth, which have three points. Parore appears to be a fish name unique to New Zealand and is not used for other fish in the Pacific. It is not certain exactly what it means, but one suggestion is 'gentle, soft or agreeable'.

The fish has been given many English names: mangrove fish, black fish, black perch, black snapper, and sometimes even 'trash fish' or 's**t fish'. Mullet fishermen who had to pull the spiky fins out of their nets called them 'sweet briars', a reference to the thorny sweet briar rose.

BIOLOGY

Parore are found in coastal waters around estuaries and rocky reefs, their striped patterning helping them to blend in among tall stalks of kelp. They primarily eat seaweed, especially green seaweeds like sea lettuce (*Ulva lattuca*) and grass kelp (*Ulva intestinalis*), but will opportunistically eat small crustaceans, mussels, pipi and worms. This diet may be partly responsible for their bad reputation as an eating fish, as sea lettuce has high concentrations of bromophenols, which are known to produce an 'off' flavour.

CHANGING ATTITUDES

There are some signs that New Zealanders' attitudes to parore are beginning to change. They are becoming a more popular fish to eat, and in places where they are highly abundant could provide an alternative to the overfished snapper. Parore is often remembered fondly as one of the first fish caught by children when fishing in harbours with a burley of green seaweed or grass clippings.

In Katikati near Tauranga there is even a conservation project that uses parore as its mascot. Run-off from the neighbouring land has led to prolific blooms of sea lettuce, one of the favourite foods of parore. The group – Project Parore – aims to restore the local harbour environment to a point where parore numbers increase and can help graze the abundant sea lettuce.

Though they are sometimes viewed with disgust in New Zealand, parore have a charming and endearing nature. *(Sophie Journee, www.emr.org.nz)*

Butterfish / Rarī

Sneaky vegetarians

Butterfish (*Odax pullus*) by Edgar Waite (1911), adapted by Lars Quickfall.

Butterfish are arguably among the most beautiful fish in New Zealand waters, gliding gracefully through forests of kelp, their rippling fins like blades of seaweed. They have a very peaceful nature, only nibbling small bites out of their favourite brown seaweeds. Yet despite being completely harmless, they have often been viewed with suspicion. For Māori there was a feeling that butterfish or rarī were troublemakers, and the word rarī became shorthand for someone who was up to no good. They were often linked with nocturnal mischief, as in the whakataukī 'Ka pō, ka pō, ka kai te rarī' ('When it is night, the butterfish feed').

Despite this, they were regularly eaten across the country, especially in southern New Zealand, where they are more common and grow larger. Their preferred habitat, around kelp forests, required specialist fishing techniques, and they were usually caught by Māori using large pole nets with bits of seaweed woven into the edges for camouflage. Another technique was to fasten a net across a gap in a rocky reef where the water would rush in and out, sweeping large numbers of butterfish into the net as the tide receded.

A butterfish swimming through stands of kelp. *(Daan Hoffman)*

GREENBONES

Pākehā, too, have been very suspicious of the gentle butterfish. There was a strong prejudice against them due to their slimy black appearance and green bones. Writing in 1872 in a book about edible fishes of New Zealand, the scientist James Hector described them as good food, but admitted they had 'a rather forbidding appearance … dark coloured slimy skin and inelegant form'.[15] Considerable quantities were buried or disposed of by fish merchants when they couldn't be sold. On more than one occasion butterfish bought at market were thrown out, with customers complaining that the bones had turned green with rot. However, this green colour is actually a feature of the fish, and is a consequence of their kelpy diet.

A turning point in the relationship with butterfish was the discovery that they were a potent source of iodine. New Zealand soils are naturally deficient in iodine – a critical

TAXONOMY

There are two species of butterfish or rarī in New Zealand. The most common butterfish (*Odax pullus*) is found throughout New Zealand, but is more abundant and larger in the South Island. In Manawatāwhi/Three Kings Islands it is replaced by the blue-finned butterfish (*O. cyanoallix*). Juvenile butterfish were once considered so different from the adults that they were classed as their own separate species, *O. vitattus*, a name that is no longer used.

ETYMOLOGY

The name *Odax* means 'biting', as the front margin of the fish's jaw forms a beak for cutting chunks out of kelp, and *pullus* means 'dark-coloured'. The Māori names 'mararī' and 'rarī' originated in the Pacific, where similar words are used to refer to relatives in the wrasse and parrotfish family. The butterfish was often called 'greenbone' in the South Island, on account of its striking blue-green bones.

Juvenile butterfish were once known as 'kelpfish', although today this name is more commonly used for hiwihiwi (*Chironemus marmoratus*). Where the name 'butterfish' came from isn't quite clear – it may originally have been a reference to the yellow colour of the juveniles, but today it is widely considered a statement on its delicious taste.

BIOLOGY

Butterfish are found throughout New Zealand on shallow, rocky reefs with lots of kelp. They are picky eaters, nibbling neat little circles out of brown kelp or picking off the reproductive parts of seaweed. Collectively, their feeding can play a major role in shaping the seaweed community of the reef.

Butterfish undergo remarkable transformations throughout their lives. They begin their lives as golden-yellow females, then they have a dark green-blue colour phase before maturing into a browny-yellow adult. After they reach around 40 centimetres in length, about half will transition into males and become a striking bright blue colour.

The yellow colour of the juvenile butterfish provides excellent camouflage in kelp. *(Daan Hoffman)*

element for growth – and in the early twentieth century there were local epidemics of goitre in humans caused by iodine deficiency, which causes the thyroid gland to swell. A public health campaign was launched in the 1920s to test for natural sources of iodine. While seaweeds generally have extremely high quantities of iodine, convincing a sceptical public to eat them was challenging. Butterfish, however, with their diet of seaweed, turned out to be the richest source of iodine of all the New Zealand animals tested. So the Department of Health began to promote the nutritional benefits of eating butterfish, and Professor Charles Hercus of the Otago Medical School recommended eating it twice a week.

GOING BUTTERFISHING

Fishermen began to have a change of attitude to the fish as well, realising that butterfish are surprisingly good fighters when caught on rod and line. Some even argued that 'butterfishing' was a more enjoyable sport than trout fishing. On calm summer days in the 1920s, up to 150 fishing boats could be observed around Shag Point on the Otago coast, fishing for butterfish, and David Graham of the Portobello Aquarium claimed it was perhaps the most sought-after fish among sports fishers in the 1950s.

Part of the appeal was that a great deal of skill and cunning was needed to persuade butterfish to bite. Fishermen needed to come armed with an array of hooks and baits, and adapt to whatever mood the butterfish were in that day. Despite being mostly vegetarian, butterfish will on occasion snap at animal food if it drifts past them, and fishers have been known to catch them using worms, crayfish, mussels, sandhoppers and slaters. Some fishers have even caught them using flowers like daisies.

The dazzling blue colours of an adult male. *(Daan Hoffman)*

Today, angling for butterfish is no longer very common, but they remain a popular target for spear fishers. The fish can be easy to spook, so spear fishers will descend into the kelp and avoid making eye contact with the fish, waiting until their natural curiosity gets the better of them before taking a shot. Butterfish is still considered one of the best-tasting fishes in New Zealand, providing clean, creamy-white fillets that soak up other flavours well.

Blue cod / Rāwaru

Afraid of thunder

Blue cod are grumpy-looking fish, their thick lips permanently set in a moody pout. When a diver enters their territory, they may angrily chase after them and nibble at their gear or fingers. But perhaps their foul mood is partly driven by hunger, as once they have eaten they can be quite friendly, and will feed from a diver's hand and even allow their heads to be stroked. They have an incredible appetite and will nibble off the legs of live crayfish, pry open kina, and even munch through bony seahorses when they find them. They are particularly fond of pāua, and will wait for starfish to pull them away from rocks before feeding on the soft underbelly of the shellfish.

Blue cod have a tremendous appetite and will feed on a wide range of marine species. *(Ian Skipworth)*

LUXURY OF THE SEA

TAXONOMY

While blue cod (*Parapercis colias*) is certainly blue, it is not a cod but a member of the sand perch family Pinguipedidae. It is endemic to New Zealand and found throughout the country, though it is more abundant around the South Island.

Blue cod or rāwaru were a valued food fish for Māori in the South Island, where the species is much more abundant than in the north. Before Māori returned to land after a fishing trip, they sometimes removed the heads of blue cod and threw them back in the water as an offering to Tangaroa, to ensure safe passage.

The first Europeans to explore New Zealand's coast immediately fell in love with blue cod. On James Cook's second voyage to New Zealand, in the autumn of 1773, the crew arrived in Dusky Sound after spending more than 120 days away from land. Their first meal of blue cod and other fish in the harbour was regarded by many as the most delicious they had had in their lives. Cook himself wrote in his journal: 'What Dusky Bay most abounds with is fish … Some are superior, and in particular the cole fish [blue cod], as we called it, which is both larger and finer flavoured than

Blue cod (*Parapercis colias*) by Frank Edward Clarke (1871), adapted by Lars Quickfall.

any I have seen before, and was, in the opinion of most on board, the highest luxury the sea afforded us.'[16] The crew particularly liked that they could eat this fish in large quantities without the taste becoming too overpowering. For the next seven weeks they stayed in Dusky Sound eating blue cod every day – boiled, fried, smoked and baked in pies and pasties, and made into chowders, stews and soups.

SNAPPER OF THE SOUTH

Blue cod also became an important part of the lives of European settlers, especially in the south. When the fishing was good, one person with a line could bring in fifty cod in an hour. The politician Walter Pearson observed tremendous numbers around Stewart Island in 1871, writing: 'I have seen them pulled up with lines three or four to each, as rapidly as the baits could be fixed and let down. I believe four good fishermen could fill a whaleboat in three or four hours.'[17] They were so common in the Chatham Islands that they could be caught by hand. Pāua were collected from rock pools at low tide and the meat placed on rocks just out of reach of the lapping water. The hungry blue cod would attempt to leap out of the water to get the pāua, whereupon the fishers would grab them and fling them ashore.

ETYMOLOGY

The scientific name *Parapercis* means 'near percis' – a related group of fishes – and *colias* most likely refers to the name 'colefish', coined by the crew on Captain Cook's second voyage to New Zealand. The original meaning of the fish's Māori name, 'rāwaru', is thought to be 'to throw up', as blue cod sometimes regurgitate their food when they are caught and brought into a boat. Other Māori names include pakirikiri, patutuki and kopukopu, the last of which could be a reference to a full, bloated stomach.

The brown colour phase of a female blue cod. *(Daan Hoffman)*.

While blue cod was widely eaten and admired by the New Zealand public, it was even more desired in Australia, where it was said no fish could match it. In the 1910s, New Zealand blue cod could be found for sale in every major town of Australia, and the hunger for this fish helped grow the fishing industry in Southland. However, this huge demand at home and abroad led some people to emphasise quantity over quality, and, before rules were brought in, tiny blue cod were often sold, some weighing only 100 grams.

Blue cod remains an important commercial fishery today, with the fish being caught on lines and with baited pots. There are currently plans to start an aquaculture industry around the fish, as scientists have recently demonstrated that they can be bred in the lab.

Blue cod spend a lot of time on the sea floor, resting on their pelvic fins. *(Daan Hoffman).*

THUNDERSTRUCK

Blue cod were once used to predict when storms were coming. *(Helen Kettles, DOC)*

Blue cod have been observed showing unusual behaviour in response to thunderstorms. In the 1950s, David Graham, the manager of the Portobello Aquarium in Dunedin, noticed the fish acting strangely one evening, nervously darting about the tank and quivering their fins in anticipation. Suddenly there was a clap of thunder and the fish began leaping out of the tank in fright. Graham rescued them and returned them to the tank, only for them to leap out of the water again at the next thunderclap. One fish he ended up rescuing eleven times. He stayed till midnight, repeatedly returning fish to their tanks, and when he went back to work in the morning found many blue cod lying dead on the floor. Apparently this behaviour had been observed before at the aquarium, but only during storms at night.

It's not entirely clear what might cause this behaviour, but something about the thunder could trigger the blue cod's 'startle response' – a split-second reaction that fish use to burst away from their predators.

Fascinatingly, this isn't the only connection between blue cod and storms. In Ngāti Koata tradition, if stones were found in the bellies of blue cod as they were being gutted, it was seen as a sign that bad weather was on the way. It was believed that blue cod could anticipate changing weather by reading wave currents and the movements of kelp; in response, they would eat stones to weigh them down on the sea floor during big storms. There was a similar belief among Pākehā fishermen that when pebbles were found in the belly of blue cod, the fish would soon depart from the area. Many fishers today note that blue cod are almost impossible to find after a storm, and can only be caught in clear, still water.

BIOLOGY

Blue cod are found all around New Zealand but are more common south of Cook Strait. They have no swim bladder, which means they will sink to the bottom if they stop swimming. This allows them to hang about hunting on the sea floor, propped up on their pelvic fins.

They are voracious carnivores, eating mostly small fish, crabs, kina and pāua. They can grow up to 60 centimetres in length and live for up to thirty years. While they all begin life as females, some transition to males over the course of their lives.

The striking blue colour of this fish quickly fades once they are caught, becoming a darker coal black. *(Matt Ruglys).*

Snapper / Tāmure

New Zealand's most beloved fish

Snapper (*Pagrus auratus*) by Alex Murray and C. Toms (1916), adapted by Lars Quickfall.

If there is one fish in New Zealand more loved than any other, then surely it is snapper. Tributes to this fish are everywhere: they feature in sculptures, paintings and murals; the fishing section of the library is full of books and magazines about them; and more scientific research has been done on this fish than any other. They are often the sole focus of fishing trips, with New Zealanders spending millions of dollars every year trying to catch them.

Exactly what has led to this enduring love affair with snapper is hard to pinpoint, but it is probably a mix of its beauty, its fighting spirit and its delicious taste. It must also have helped its reputation that it is found in great abundance near major population centres, for example in the Hauraki Gulf, and it is a fish that can be caught by anyone, from seasoned fishing veterans to first-timers.

The beautiful blue speckles are brightest in juvenile fish and become less prominent with age. *(Daan Hoffman)*

TAXONOMY

Snapper (*Pagrus auratus*) belong to the sea bream family Sparidae. They have had over twenty scientific names over the past two centuries, as scientists have debated where to put them in the tree of life. The New Zealand species is a close relative of the common sea bream of the Atlantic (*Pagrus pagrus*) and the madai (*Pagrus major*) in Japan.

THE FOOD OF WARRIORS

Snapper was of inestimable value to Māori from the earliest days of settlement in Aotearoa, especially in the north. The archaeological record of northern New Zealand is littered with snapper bones, which account for around three-quarters of all fish bones found in middens.[18] Snapper could be caught on hook and line, but the most effective approach was to use large seine nets that stretched for over a kilometre and trapped thousands of snapper in a single haul of the net.

All parts of the fish were eaten, including the guts, eyes, heads, wings and collars, but the most highly valued part was the nene – the base of the tongue. It was a prestigious and noble food, generally reserved for the highest-ranking warriors, and was believed to make them grow active and strong. Snapper were eaten grilled, steamed, wrapped in leaves and placed on hot stones, or baked in a hāngī. One method, known as kaniwha, involved soaking the flesh in fresh water and squeezing it out several times before eating it raw. To preserve the flesh for later use, snapper were cleaned, split to the backbone and hung out to dry on wooden racks.

As such an important resource, snapper had a huge cultural impact as well. Schools of snapper were seen as a sign that all was well with the world, and they were compared with the flowering bracts of the kiekie vine on land as a symbol of the bounty that nature could provide, as in the whakataukī 'He whā tāwhara ki uta, he kiko tāmure ki tai' ('The flower bracts of the kiekie on land, the flesh of the snapper in the sea').[19]

While they were one of the favourite foods of Māori in the north of the country, according to one tale seabirds were not as fond of snapper as their human counterparts. In the story, the sharp spines of the snapper got caught in the birds' crops as they fed. As a result, the seabirds decided to wage war against the river birds and claim rights to the eel fisheries so they had a soft and slippery fish to eat. But they were soundly defeated and were forced to settle for eating the spiky snapper instead.

ETYMOLOGY

The name *Pagrus* comes from the word *phagros*, an ancient Greek name for fish in the Sparidae family, and the name *auratus* means 'golden', referring to the golden crescent that sometimes extends from one eye of this fish to the other. Māori used the name 'tāmure' for adults and 'karatī' for the juveniles. The name 'tāmure' is also given to a number of similar-looking but unrelated fish in the Pacific, such as the blacktail snapper (*Lutjanus fulvus*) and the orange-finned emperor (*Lethrinus erythracanthus*).

Fishers have developed a range of different terms for snapper at different stages. 'Pannies' are smaller fish that are a good size for cooking in a pan, 'schoolies' are younger fish that school over the sand, and 'old-man snapper' are large, reef-dwelling fish, which develop big bumps on their forehead with age.

THE SEA BREAM

Snapper made a big impression on the first Europeans in New Zealand. While fishing off Whangārei Heads in late 1769, the crew of the HMS *Endeavour* were overwhelmed by the number of snapper they caught, which they called sea bream. James Cook wrote in his journal, 'We had no sooner come to anchor than we caught between 90 and 100 bream'[20], and the immense catch led him to call the area Bream Bay.[21] As other Europeans travelled around New Zealand they frequently described throwing out their fishing lines and in no time at all catching enough snapper to feed their crew and sustain them for several days afterwards. Even a century later, in the 1880s, a fisher in Kaipara Harbour could expect to catch snapper at a rate of sixty or seventy fish an hour.[22]

And yet despite its early impact and huge abundance, snapper's status as the premier New Zealand fish was not guaranteed. For many years people disregarded it, considering it tasteless. One writer complained to a newspaper in 1868 that despite an abundance of better options 'the standard fish offered for sale is the dry, insipid and inferior snapper'[23], while the naturalist William Travers described snapper as 'poor in flavour and coarse in flesh'.[24] In some places they were referred to as a 'trash fish' and considered a pest, getting stuck in set nets for flounder and mullet and damaging them with their spines. But the prejudice eventually wore off, and snapper became a favourite of many settlers in the north of the country, who valued its sweet white flesh.

Snapper were not always regarded highly as a food fish. *(SeacologyNZ)*

For many years, fishing was seen primarily as a subsistence activity to put food on the table. But during the twentieth century more and more people got into fishing just for the sport of it, and the snapper's cunning and tenacity made it an ideal target. In those early days snapper were sometimes caught merely as trophies, with no plans to eat them. At the Doubtless Bay Campground in the early 1960s, large hauls of snapper would be brought in and laid out in front of campers' tent sites, where they sat in the sun, spoiling.[25] They could generally be bought cheaply, and were the staple fish at fish and chip shops around the country.

Today, however, they have become a much more valuable and prestigious commodity, and are more likely to be seen on the menu at a top-end restaurants, in complex dishes. New Zealand snapper is particularly sought after for sashimi and sushi in Japan – its flesh resembles the native red sea bream – and is served at New Year celebrations and at weddings and birthdays to bring good fortune.

Large adult snapper play a critical role in maintaining healthy marine ecosystems. *(Department of Conservation)*

SNAPPER FEVER

While some fish languish in obscurity, snapper is perhaps too famous for its own good. New Zealanders' love affair with snapper is so strong that these creatures are often the sole focus of fishing trips, during which many other perfectly edible fish are thrown back. Even when snapper are caught, people often eat only the thin fillets from either side of the fish, while the heads, cheeks, roe, collars and wings are discarded.[26]

This 'snapper fever' has implications for the reef, as snapper play a critical role in the ecosystem. Together with crayfish, adult snapper keep kina populations under control and prevent them mowing down kelp forests and clearing the reef of habitat for other species. Managing the harvest of snapper can help reverse the trend in some areas, allowing snapper to grow large enough to take on the spiky kina. At the fully protected Goat Island Marine Reserve north of Auckland, snapper have grown to immense sizes, with one old-man snapper, known affectionately as Monkey Face, thought to be around eighty years old.

Snapper are generalists, able to survive in a wide variety of habitats with a range of different food sources. *(SeacologyNZ)*

BIOLOGY

Snapper are found around the North Island and the top of the South Island. They are the great all-rounders of the ocean, and can live in a wide range of habitats and feed on a diverse array of sea creatures. One study found nearly 100 different species in their stomachs, including crustaceans, worms, starfish, urchins, limpets, shellfish and mussels.[27] Juvenile snapper settle in shallow estuaries and are strongly reliant on seagrass beds as habitat, to hide from predators. As they grow, they move out onto the coast and start to eat a wider range of foods. By around three to five years old they start to mature and are able to tackle larger prey items such as the spiky kina.

OCEAN HUNTERS

Roaming the open seas

Sharks / Mangō

Noble warriors

Great white shark (*Carcharodon carcharias*) by E. N. Fischer (1913), adapted by Lars Quickfall (*Harvard University*); spotted dogfish (*Mustelus lenticulatus*), spiny dogfish (*Squalus acanthias*) and school shark (*Galeorhinus galeus*) by Frank Edward Clarke (c.1870s–1880s).

Sharks have a powerful effect on the human psyche, and wherever they appear they are met by terror, disgust, awe and wonder. Perhaps it is unsurprising that sharks have such a deep emotional impact, as they are truly ancient creatures. When the first dinosaurs emerged on land, sharks had already been roaming the oceans for hundreds of millions of years. Over these unfathomable timescales, evolution has turned sharks into the ultimate ocean predators, with sleek, torpedo-shaped bodies, the ability to detect minute amounts of blood in water, and rows of teeth that are constantly replaced and kept razor sharp.

Sharks were highly revered by Māori, who especially admired their fighting spirit.[1] Their unwillingness to surrender when caught was seen as a metaphor for a warrior who never gives up, and soldiers were encouraged with the whakataukī 'Kaua e mate wheke, me mate ururoa' ('Do not die like an octopus, die like a shark').

Large sharks were often regarded as kaitiaki, spiritual guardians that protected those who got into trouble at sea, and many of the original voyaging canoes that travelled to New Zealand from Hawaiki were said to have been assisted by sharks. The voyaging waka *Ngā rākau rua a Atuamatua* was attacked by Te Parata – a terrible creature known as the mouth of the ocean – but before it could be sucked into its jaws, it was rescued by a giant shark, which towed them to safety. In honour of the creature that had saved their lives, the voyagers renamed their waka *Te Arawa*, another name for shark.

TAXONOMY

There are over sixty species of shark in New Zealand waters. The largest, most recognisable sharks around our shores are the mako (*Isurus oxyrinchus*), the bronze whaler (*Carcharhinus brachyurus*) and the infamous great white (*Carcharodon carcharias*). The species most commonly used for food in New Zealand waters are the 'dog sharks' – the spiny dogfish (*Squalus acanthias*), the spotted dogfish (*Mustelus lenticulatus*) and the school shark (*Galeorhinus galeus*), which are mostly harmless to people.

The unmistakable silhouette of a great white. (*Elias Levy*)

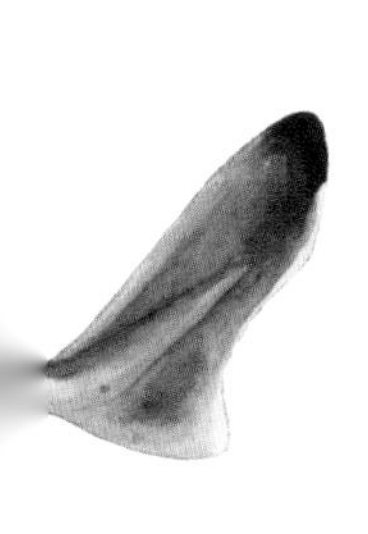

A PUNGENT DELICACY

For Māori, sharks were a vital food resource, especially the smaller dog sharks, and a diet of shark meat was believed to make people grow strong and brave. Sharks were typically caught with nets and hooks in shallow bays and harbours. To preserve the meat, sharks were hung on wooden frames, some up to 400 metres long. Once they had dried, the carcasses were stacked like logs of firewood in special storehouses dedicated solely to shark meat.

Dried shark produces a strong ammonia-like smell. One tradition explained that the stench came about because a shark had refused to help the atua Hine-te-iwaiwa in her search for her husband, Tinirau, so she urinated on it as revenge. The pungent smell was said to be worth putting up with for the taste, however, which was considered among the best seafood available and highly valued all around the country.

ETYMOLOGY

The word *Squalus* is an ancient Latin word for animals that were considered filthy and not fit for human consumption, and the names *Mustelus* and *Galeorhinus* are derived from words meaning 'weasel'. There is a lot of debate about the etymology of the word 'shark'; some point to the German word *schorck*, meaning 'a dishonest person', while others suggest it comes from the Aztec word for sharks, *xoc*. Mangō is the generic name of sharks in Māori and is used across Polynesia for shark species. The names of the supernatural creatures 'taniwha' and 'Te parata' are both derived from Polynesian names for sharks.

Dried shark was often the basis of immense feasts: an 1844 hākari in Remuera had 4000 guests and served 9000 sharks[2], and the politician William Swainson claimed that at a Waikato hui in the 1850s, 20,000 sharks were served.[3] To ensure these huge catches were sustainable, shark fisheries were strictly managed. In Raungaunu Harbour in the Far North, only two days a year were

A CATCH OF 54 SHARKS AT AWAROA BAY, WAIHEKE, AUCKLAND. MARCH 5, 1906. Jones and Coleman, Photo.

A catch of sharks drying on a wooden rack. *(Auckland Library Heritage Collections)*

set aside for fishing school shark (kapetā) and hapū from all around Northland would travel great distances to participate in these open days. The naturalist R. H. Matthews attended a harvest in 1855 and described a wild scene with fifty waka hauling in more than 7000 sharks between them.[4]

TEETH AND OILS

Food was not the only thing that sharks were useful for. The oil extracted from shark livers was an important cosmetic; it could be coloured, by mixing it with red ochre (kōkōwai), and perfumed with fragrant plants. The oil was used to anoint the body and hair, rubbed on newly born babies and even used as a healing balm on cuts and wounds. It helped to preserve and colour woodwork and was painted over canoes, houses, marae buildings and carvings. It had an especially important ceremonial role in funerals, and was used to anoint the bodies and bones of the deceased and adorn carved monuments at urupā. This made the oil a particularly valuable commodity and often a key item of trade, especially with inland iwi who lived far from the coast. It was later sold to Europeans to be used as lamp oil.

The teeth of larger sharks such as mako and great white were considered beautiful and were used for necklaces and ear pendants.[5] They had practical uses as well, and were made into battle clubs and knives, and used for cutting hair, shaving beards or scarring the skin during mourning ceremonies.[6] Collecting the teeth, while preserving these revered species, required great courage and skill. Māori fishermen would use a stingray as bait to lure in a shark, then they would lasso it around the tail or gills with a strong rope and let it tow their waka until it tired out. Once they had exhausted the shark, they would carefully extract the teeth (which would soon regrow), before setting it free.

BIOLOGY

There is huge diversity in shark breeding patterns. Some sharks, such as the catsharks, lay leathery egg cases; others, like great whites, keep their eggs inside the uterus, which hatch there and continue to develop until live babies are born; and yet others, like mako and hammerhead, give birth to live young. One of the only observations of great whites mating was recorded around Dunedin, when two giant sharks locked together and began rolling around and around in the water.

Many sharks require water to be constantly moved over their gills in order to oxygenate their blood, and so must always keep swimming.

CREATURE OF FEAR

Pākehā had more of a mixed relationship with sharks. Some caught the smaller dog sharks to eat or sell at market. They used their rough skin as sandpaper, and even rubbed shark liver oil into cattle to help relieve them of parasites. But many viewed sharks as a terrible pest that would ruin nets and fishing lines. In the early twentieth century, flounder fishers in Kaipara Harbour would catch so many spotted dogfish in their nets that they took to breaking their backs and dumping them in the water. There was little appetite among the public for eating sharks, so fishmongers tended to sell them using other names that were more palatable, such as 'silver strip' and 'lemon fish'. Under these names they were generally well-liked and sold better than many other fish, and they remain a common item in fish and chip shops today.

OXYRHINA GOMPHODON.

The shortfin mako (*Isurus oxyrinchus*) is one of the fastest fish on the planet and is known for leaping to great heights out of the water when hunting. Artwork by Johannes Müller, 1841. *(Harvard University)*

A great white. Sharks can be dangerous and deserve respect, but are not the monsters they are sometimes made out to be. *(SeacologyNZ)*

Pākehā also came to admire the ferocious tenacity of the larger sharks, and over the years they became popular hunting prizes that were often hauled into towns and exhibited.[7] In the 1920s, American writer Zane Grey was invited to New Zealand in the hope that he would promote the country's fishing. He was particularly enamoured of the mako sharks he observed here, describing them as 'an engine of destruction, developed to the nth degree'.[8] His reports spurred a surge of tourist interest in New Zealand and ultimately resulted in the word 'mako' being adopted for that species of shark around the globe, making it among the most widely used Māori words in the world.

Today, fear of sharks is still common, driven in no small part by Steven Spielberg's monster film *Jaws*, which led to a huge decrease in beach attendance for many years after. But while sharks can certainly be dangerous to people, over the past 170 years there have been only thirteen fatal shark attacks in New Zealand.

Sharks have far more to fear from us than we have from them. Over the past fifty years alone, nearly three-quarters of the global shark population has been wiped out, and the fishing pressure on sharks remains intense, with these creatures often being killed solely for their fins. However, there are some glimmers of hope in our relationship with sharks, and more people are fighting to protect them. There is a much greater awareness of the important role they play in maintaining healthy ecosystems, and great white sharks are now fully protected in New Zealand waters and cannot be injured or killed, under threat of huge fines or jail time.

The smooth hammerhead (*Sphyrna zygaena*) by Marc Éliéser Bloch, 1796. *(Harvard University)*

Kahawai

The people's fish

Kahawai (*Arripis trutta*) by Frank Edward Clarke (1872), adapted by Lars Quickfall. *(Te Papa)*

Kahawai are active hunters. The smallest sign of movement in the water is all they need to begin pursuit, thrusting after prey with their large, powerful tails. Māori caught them on trolling hooks without any bait, using shiny pāua shells as a flashy lure to attract their attention. The selection of the pāua was very important, and shells flecked with red were thought to be irresistible to kahawai.

The best time to catch kahawai was in summer, when they swarmed the river mouths to spawn. These runs of kahawai were valued so highly that communities of hundreds of people would leave their permanent homes and camp out at a river mouth for several months, focusing solely on catching and preserving kahawai. These seasons could be exciting social events, with more than twenty waka powering up and down the river at full speed, trolling pāua lures behind them, while the women jumped and shouted from the shoreline and the men sang in unison.

Using this method, fishers could catch 200 to 300 kahawai each on the incoming tide. Once caught, the kahawai were brought back to shore, scaled, cleaned and steamed in a hāngī. All parts of the fish were used, including the head, throat, roe and other organs. Those fish not eaten right away were preserved to provide a valuable winter food source, by being hung up on racks to dry and harden.

TAXONOMY

Kahawai (*Arripis trutta*) are in the Arripidae family, a group that is found only in New Zealand and Australia. It is easily confused with the Kermadec kahawai (*A. xylabion*), which also occasionally occurs in northern New Zealand waters, but this species has a larger tail.

A pā kahawai – kahawai trolling hook – made from a colourful pāua (*Haliotis iris*) shell. *(Te Papa, OL000106/10)*

A SOURCE OF CONFLICT

Kahawai are particularly 'bloody' fish and need to be bled as soon as they are caught or the fillets will soon become stained with blood. Because of this, kahawai were often compared with women, who suffered a lot of blood loss during childbirth. It was said that 'He ika toto nui: he kahawai ki te moana, he wahine ki uta' ('A much-bloodied fish: the kahawai in the sea and the woman on land'). It was perhaps this connection that meant in some areas women were not permitted to partake in the kahawai harvest.

Kahawai were also the source of much bloodshed between people, and battles over kahawai fishing grounds were not uncommon. Possibly the largest land battle fought on New Zealand soil was started over a kahawai harvest at the Marokopa River on the west coast of the North Island in the early 1800s. Conflict broke out over claims the kahawai had not been evenly distributed, and the rangatira Pikauterangi was nearly drowned. Dozens of iwi were drawn into the conflict, which culminated in a pitched battle at Lake Ngaroto, near Te Awamutu. The fight ultimately claimed thousands of lives and was named the Battle of Hingakākā (falling kākā) for all of the rangatira who perished in the fighting.

THE REVENGE OF POU

Perhaps nowhere were kahawai quite as important as on the Motu River in the eastern Bay of Plenty, which was believed to be the source of the species. In one tradition, the huge abundance of kahawai in the area was the result of a bitter dispute with Tangaroa. The local rangatira, Pou, had a son who had drowned in the river, and he blamed Tangaroa for the death. So Pou decided to get revenge, and invited the atua to come to his son's tangi. When Tangaroa approached the land, escorted by thousands of kahawai, Pou made the signal and his people ran into the water with a great net, catching huge numbers of kahawai.

Today kahawai are still an important taonga in the area, and kahawai symbols are seen in local murals, carvings and pou, and form the emblem of the local preschool. There are kahawai-inspired kapa haka performances, and local schoolchildren learn how to make pāua lures in the traditional way.

ETYMOLOGY

The Latin name *Arripis* means 'without fans' and refers to the patterns on this fish's scales, and the name 'trutta' means 'trout'. Kahawai is one of the few fish species in New Zealand for which the Māori word is the most commonly used name; however, it is frequently mispronounced as 'kawai'. The name means 'strong in the water' and must have originated in New Zealand, as it is not used elsewhere in the Pacific. It could relate either to the species' strength in fighting when caught or its preference for strong currents.

A RIVER OF FISH

Early Pākehā settlers reported seeing immense, unbroken shoals of kahawai around the New Zealand coast that stretched from horizon to horizon. When the kahawai ran in the Wairau River in Marlborough, it was said the placid waters began to bubble and froth as if they were being tossed by a strong wind. A fisherman and his children netting kahawai on the Kaupokonui River in 1927 were overwhelmed by so many fish pouring into the river that they were knocked off their feet. The fisherman described the incredible scene: 'It was literally a river of fish. The fins and tails showed above the water, so close as to almost touch one another. Away out to sea, even through the rough breakers, the stream of kahawai continued, all trying to enter the stream but prevented by the congestion of those in front.'[9]

Pākehā admired the strong fighting spirit of the kahawai when caught, but were unimpressed by the fish's taste, describing it as dry, coarse and flavourless. Australian biologist Theodore Roughley, in a 1916 book celebrating the values of edible fish species, described them as being of 'rather inferior quality' and, strangely, claimed they were not a good fish for smoking.[10]

Pākehā prejudice against the fish lasted in New Zealand for a long time, with some fishers refusing to take kahawai when caught, even when no other fish were taken that day. Part of this dislike for kahawai may have resulted from catching larger specimens, which can have a drier flesh, and not properly bleeding the fish after capture.

Kahawai travel in huge roaming schools of fish. *(Daan Hoffman)*

A NATIONAL DELICACY

Today, however, most of this prejudice has disappeared, and kahawai are one of the most commonly caught recreational species – second only to snapper. While many fish have to be caught from a boat, kahawai are accessible to everyone, and can be fished for anywhere around the coast of New Zealand. As a result, kahawai have come to be symbolic of the right of every New Zealander to fish, and are known as 'the people's fish'.

Their fighting strength when hooked has earned them a reputation as one of the best saltwater fly-fishing species in the world. When properly prepared, kahawai are a beautiful-tasting fish, and smoked kahawai is widely seen as a New Zealand delicacy. The species is served at a number of top restaurants: chef Monique Fiso serves it at her Wellington restaurant, Hiakai, cured with the bark of the native manono tree (*Coprosma grandifolia*).

BIOLOGY

Kahawai are found throughout New Zealand but are most common in the North Island. They are a pelagic fish – living in the open ocean – and travel in huge schools, with the largest exceeding a million fish. Kahawai are fast-moving and fast-growing, and will migrate hundreds of kilometres around the coast. When feeding, they force schools of small fish up to the surface, where they are often eaten by white-fronted terns (*Sterna striata*). The presence of these 'kahawai birds' is sometimes used by fishers as a way of finding schools of kahawai.

White-fronted terns (*Sterna striata*) are known to anglers as kahawai birds, as their presence can indicate schools of feeding kahawai. Artwork by Joseph Wolf, c.1839. *(Alexander Turnbull Library)*

Kahawai are now highly admired for their fishing and eating qualities. *(S. W. Geange)*

Kingfish / Haku

Chief among fish

Kingfish (*Seriola lalandi*) by Arthur Bartholomew (1885), adapted by Lars Quickfall. *(Smithsonian Libraries)*

Therc is perhaps no more awe-inspiring sight in New Zealand seas than a swirling vortex of kingfish. These roaming schools of predators often appear seemingly out of nowhere, then circle in a towering whirlpool before disappearing again into the gloom. They are powerful and effective hunters, surprising fish by ramming into them at full speed, and sometimes driving themselves onto beaches in pursuit of their prey. They are such voracious hunters they will even attack birds rafting at the ocean surface. One large kingfish caught near the Mokohinau Islands off Northland was found to have eaten four fairy prions, while another near White Island in the Bay of Plenty had swallowed a diving petrel and a pair of red-billed seagulls.[11]

Despite being a top predator, kingfish are surprisingly curious fish. They have excellent hearing, and will immediately investigate any strange noise in their environment. They are often drawn to boat noise or the sound of a diver's exhaust, appearing suddenly and disappearing again once their curiosity has been satisfied. Spear fishers will sometimes make grunting noises or strum on their spear guns to get their attention. If they succeed in shooting one, more kingfish are often drawn to the area to see what the commotion is about.

A whirlpool of circling kingfish is a majestic sight for any diver. *(Ian Skipworth)*

TE IKA RANGATIRA

Kingfish were regarded by Māori as among the most delicious fish in Aotearoa. They were generally caught without bait, using hooks made with shiny pāua lures. They could also be swept up in large nets off the beach, and a lot of strength was needed to pull in a school of them.[12] Kingfish made excellent trading items or gifts, and fishermen would sometimes carry these great fish a long distance inland to share with others.

Māori especially appreciated the tenacious energy of the kingfish, and compared them to a warrior or leader driving enemies away. It might be said of a great rangatira with enormous determination: 'Kitea te ānga nā te haku!' ('See the driving force of the kingfish!')

In some parts of the country kingfish were considered a sacred fish, and associated with the end of life. If a great rangatira suddenly started to hunger for the taste of kingfish, it was seen as a premonition that he might die soon.

TAXONOMY

Kingfish (*Seriola lalandi*) are part of the jack family Carangidae, which includes trevally (*Pseudocaranx georgianus*) and jack mackerel (*Trachurus novaezelandiae*). They are found mainly around the North Island but also occur throughout the southern hemisphere.

CHASING KINGIES

In a familiar story for New Zealand sea creatures, kingfish developed a poor reputation among European settlers, who generally regarded it as coarse and flavourless.[13] However, those people who gave the fish a chance tended to like it, especially once it was canned. Some old recipes recommend curing and smoking the belly, or boiling the flesh and serving it with egg sauce.

Whakataukī link the powerful movements of kingfish in the water with the determination of a decisive leader. *(SeacologyNZ)*

Although there were mixed reviews of its taste, kingfish quickly endeared itself to settlers in New Zealand as a sporting fish. Today, New Zealand is generally considered to have one of the best kingfish fisheries in the world, and the species tends to grow bigger here, attaining far more fishing records than anywhere else. The overwhelming majority of kingfish are caught by recreational fishers, and they are one of the most highly prized fishing targets in the country, with a large number of charter operators offering people the chance to catch one.

SUSHI KING

As with many of our fish, there has been a complete reversal in attitudes to kingfish, and they are now highly regarded as a seafood and can command huge prices at market. They provide thick, firm fillets that can be cooked like steaks.

To help meet the growing demand for kingfish, researchers at the National Institute of Water and Atmospheric Research (NIWA) have worked out how to take these kings of the sea and farm them on land. The fish are reared through their entire life-cycle from egg to adult in purpose-built tanks in which salt water is constantly circulated. The females release hundreds of thousands of eggs, which are fertilised by the males and float to the surface. Here they are scooped up by scientists, checked for quality and incubated. The tiny fish that hatch are fed on a diet of aquatic animals and formulated meal, and in one year they are ready to harvest.

Farmed New Zealand kingfish has now become a premium food in its own right. It has received a number of taste and food awards, and is served in restaurants around New Zealand and the world. Compared with wild kingfish, it is said to be much richer and creamier, with a sweet, light flavour, and it is especially sought after for sashimi and sushi.

Juvenile kingfish are farmed in tanks for sale in restaurants around New Zealand. *(SeacologyNZ)*

RAY RIDERS

While kingfish are effective hunters in their own right, they are perfectly happy to team up with other sea creatures as well. In shallow harbours and estuaries, kingfish will often swim with stingrays, coasting above or just behind them. Stingrays have a useful ability that kingfish lack, a series of electroreceptor pores around their mouth, which they can use to hunt out the electrical signals produced by fish and crabs. The kingfish use the stingrays as floating fish-detectors, and when flounder and other fish are disturbed by the stingrays' wings, they quickly dart after them and snap them up. This has proven to be such an effective strategy that a single stingray can sometimes be seen being followed by a train of five or six kingfish.[14]

This behaviour has led to an increase in the popularity of saltwater fly-fishing for kingfish in estuaries. Fishermen first seek out stingrays with an entourage of kingfish and target them with fly lures, sometimes catching these great ocean predators in knee-deep water.

ETYMOLOGY

It's not entirely clear why the name *Seriola* was given to this group of fish, as it comes from the word *seria* meaning a large pot. The name *lalandi* honours the French naturalist Pierre Antoine Delalande, who collected specimens for the scientist who first described this fish. The Māori name 'haku' doesn't have a clear origin either, and may possibly be derived from the Polynesian words *saku* or *saku-laa*, which are used to describe swordfish, sawfish, sailfish and marlin. The name 'kingfish' is given to a number of large, impressive fish in the jack family, and they are often affectionately known as 'kingies'.

Despite being top predators, kingfish will happily trail after stingray to scavenge a free feed. *(SeacologyNZ)*

Juvenile kingfish are almost unrecognisable. Their golden-yellow stripes help camouflage them among floating beds of kelp. *(SeacologyNZ)*

BIOLOGY

Kingfish usually live in schools of up to 100 fish. They swim in the open ocean but are drawn to areas of strong current around rocky outcrops, reefs and pinnacles, where they feed on schooling fish such as piper, blue maomao, trevally and kōheru. When they are young, they are themselves prey to trevally and kōheru. As adults, their silvery skin absorbs and reflects polarised light, making them appear almost invisible in the open ocean.

Barracouta / Mangā

A voracious predator

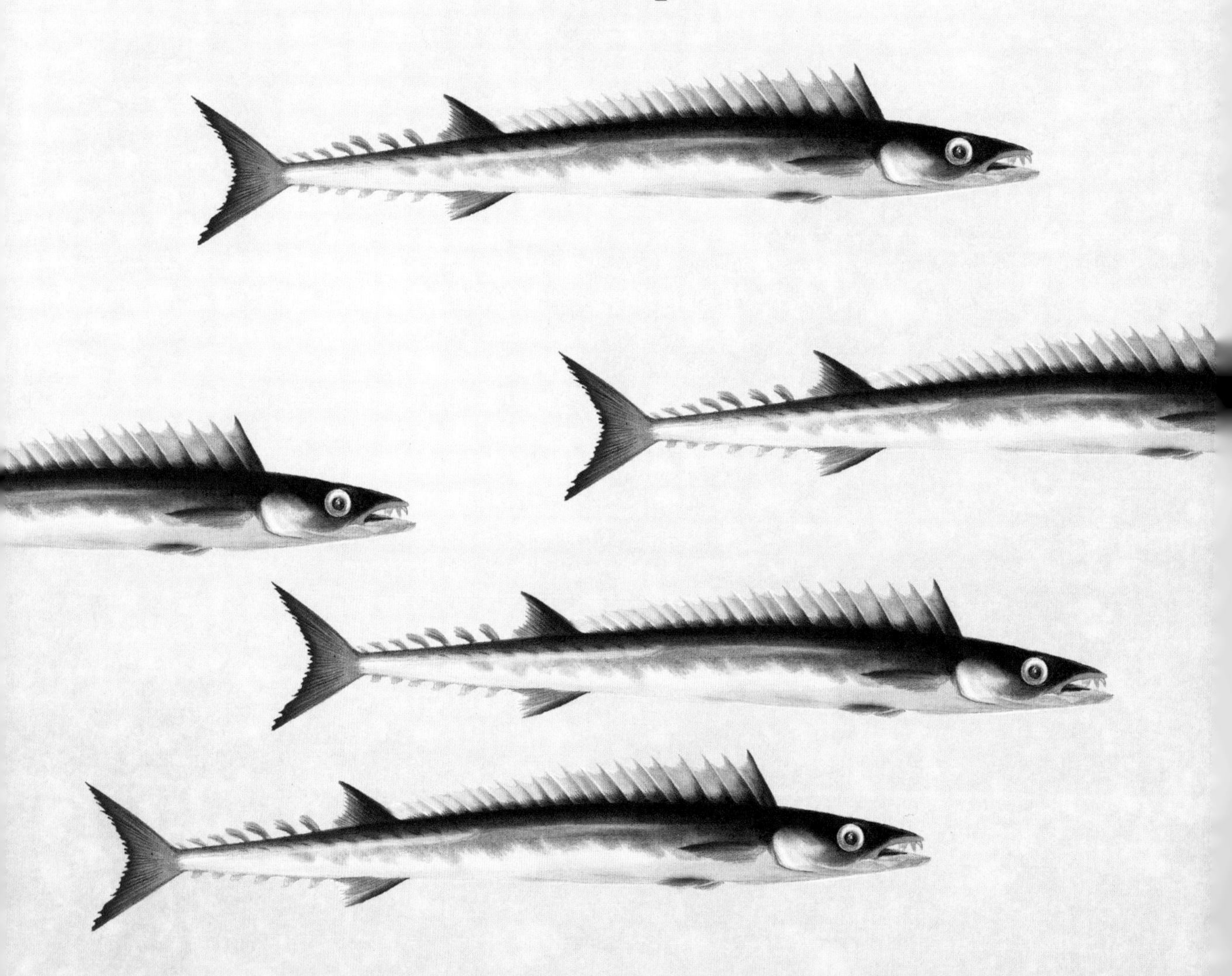

Barracouta (*Thyrsites atun*) by William Below Gould (c.1832), adapted by Lars Quickfall.

Barracouta work themselves into such a feeding frenzy that they will bite at just about anything moving in the water. Māori fishers would catch them without bait by using pā mangā – lures comprising a piece of red wood from a southern beech tree and a hook made from bird bones or dogs' teeth. Once they had found a school of barracouta, the fishers would lower these lures into the water and thrash them about vigorously to mimic the movements of small fish. Such a target proved impossible for the barracouta to resist and again and again they would lunge for the bait as soon as it hit the water.

Once the barracouta began to bite, the fishing was non-stop. In one fluid motion, fishers would catch a barracouta, flick it over their shoulder and immediately resume fishing. With several fishers attacking the water like this, there could be an almost unending stream of barracouta being pulled from the ocean and flung into the air. One European observer witnessed Māori fishing for barracouta at Stewart Island in 1870 and described the incredible scene: 'One by one, as swiftly as the rod can be wielded, the lithe forms drop off the barbless hook into the boat, till the vigorous arm can no longer respond to the will of the fisherman, or the vessel will hold no more.'[15]

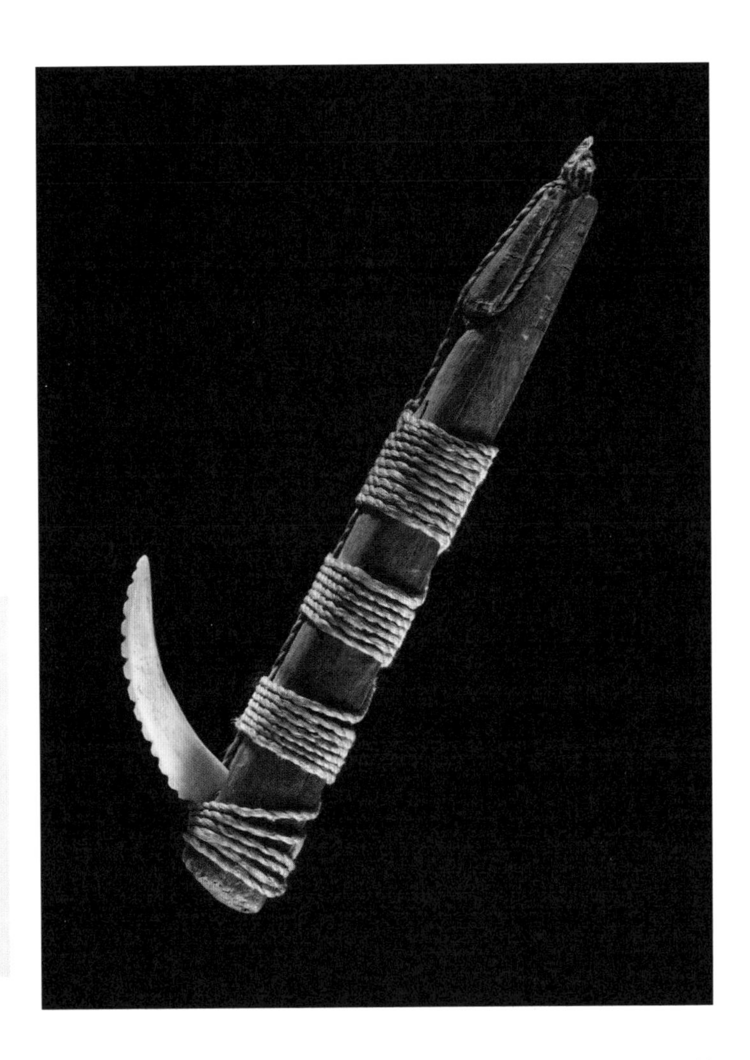

A pohau manga – barracouta lure – made from wood, bone and flax fibre. This hook is thought to have been collected by Captain James Cook during his voyages to New Zealand. *(Te Papa, ME002494)*

TAXONOMY

Barracouta (*Thyrsites atun*) belongs to the snake mackerel family (Gempylidae). Barracouta are found throughout New Zealand but are more common in the cooler waters south of Cook Strait. They are widespread across the southern hemisphere and are also found in South Africa, South America and Australia.

In the colder south of the country, where the climate made growing crops difficult, barracouta was one of the most important food sources for Māori, and they were caught in far greater abundance than any other fish. A number of major settlements in the South Island are thought to have been based around access to good barracouta fishing grounds, and communities would often move great distances around the landscape and camp seasonally to fish for barracouta. While it was men who were typically involved with catching the fish, women focused on processing them: cutting them open, washing them in fresh water and drying them on wooden racks in the sun. To cook the fish, enormous hāngī were constructed and the barracouta slow-roasted over several days. Once cooked, the flesh was scraped from the bones and stored in a pōhā – a bag made of bull kelp seaweed, sealed with fat. If done correctly, these kelp bags of preserved barracouta could last several years and became a valuable emergency food source.

ETYMOLOGY

The scientific name *Thyrsites* means 'stalk-like' and refers to the barracouta's long body; *atun* comes from the Spanish word for tuna, *atún*.

The Māori name 'mangā' is pronounced *makā* in the South Island. The name 'mangā' is used for similar fish in the Pacific, such as the singleline gemfish (*Promethichthys prometheus*) in Rarotonga, which is in the same family and looks alike. The English name 'barracouta' comes from the fish's similarity to the barracuda (*Sphyraena sphyraena*).

Although they were valuable, barracouta could be a pest when Māori were trying to fish for other species, as they would get into fishing nets and tear them apart with their sharp teeth. As a result, a person that got in the way or ruined a well-laid plan was sometimes compared to a barracouta feeding in the sea. Barracouta were known to disrupt the plans of the atua as well. In one tale, Tangaroa gave the barracouta an important job, but it got distracted when it saw a shoal of fish and chased after them instead. As punishment, Tangaroa made the bones of the barracouta sharp like a spear so that they pierced the inside of its body – which explains why the flesh of barracouta are full of sharp bones that can be a challenge to avoid when eating the fish. The atua of death and the underworld, Hine-nui-te-pō, was described as having the mouth of a barracouta, filled with sharp fangs.

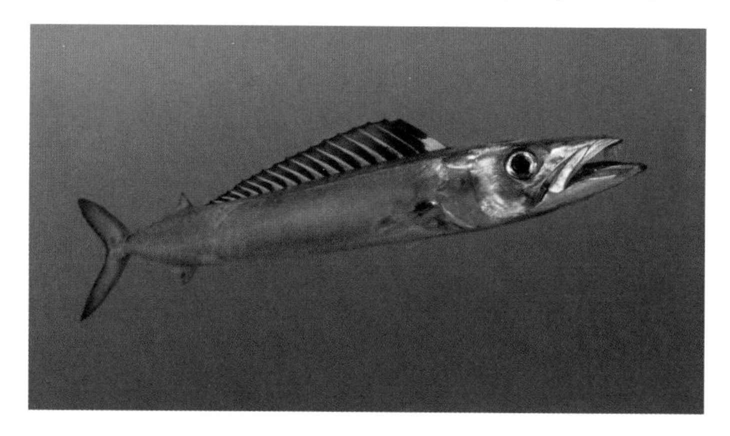

Barracouta were once a vital food source, especially in the South Island, where it was harder to grow crops. *(www.gnomad.org)*

BIOLOGY

Barracouta are found throughout the southern hemisphere and generally prefer cooler waters. They will travel long distances to reach their spawning grounds. They can grow up to 1.5 metres long, with the largest fish recorded measuring up to 2 metres in length and weighing over 6 kilograms. In late summer and autumn, large mature fish come close to shore to feed at the surface on sprats and pilchards, and this is when they were most commonly targeted by Māori.

THE COOTA STICK

During the early years of European settlement in Otago, settlers relied heavily on barracouta caught by Māori to help feed the growing colony. As time went on, European fishers began to adopt the Māori technique of barracouta fishing, using a piece of wood with a bent nail that they called the 'coota stick'. Some fishermen became highly proficient at its use; in the 1880s it was said that at Otago Heads two men fishing in a row boat could catch around 500 fish in several hours[16], and the there is even a report of a fisherman catching over 1100 fish in a single day.[17] The record for a single day's fishing by a fishing vessel operating out of Port Chalmers was over 8500 fish. To process these incredible numbers of fish, fishers developed a method for splitting them in half and removing the spine in under twenty seconds.

Barracouta (*Thyrsites atun*) by Jacques Christophe Werner, c.1830s.

PET FOOD

The meat of barracouta was generally viewed as very coarse, but early Pākehā settlers were happy to eat it anyway, as it was cheap and abundant. It became a common food for gold-diggers in the South Island, where it was available at any boarding house.

Today, barracouta are not especially popular as food, as their bones make them difficult to eat and the flesh is sometimes full of white parasitic worms. When caught, barracouta is often used as bait for other more desirable species such as crayfish, or made into pet food.

But for those who develop a taste for them, they can be eaten grilled, fried, smoked, pickled or dried. Barracouta is a favourite of the South African community in New Zealand, as the same fish occurs off southern Africa, where it is known as snoek and is a popular culinary delight. A large portion of the New Zealand barracouta catch is exported to South Africa and sold there.

Here barracouta remain something of a nuisance for fishermen, munching through fishing lines, eating hooked fish and tearing big holes in nets, and need to be handled with care, as their sharp fangs can cause deep cuts.

The sleek, shiny body of a barracouta powering through the water at night. *(iStock)*

John Dory / Kuparu

The miracle fish

John Dory (*Zeus faber*) by Edward Donovan, c.1808. *(Biodiversity Heritage Library)*

John Dory haunt the reef like a spectre, drifting silently through the water, their eyes rotating in their sockets to find their next prey. Once they locate their victim, they turn to face them, their thin, super-compressed bodies making them almost invisible. When their prey moves close enough, they strike with lightning-fast reflexes, their telescopic mouths shooting out and vacuuming fish into their stomachs.

John Dory are such effective hunters that they are known to gorge themselves until they are so full of fish that they can barely swim. A John Dory in British waters was once found to have three small fish, five stones and twenty five flounder in its stomach, and was so bloated it could be plucked out of the water by hand. In New Zealand, John Dory have accidentally killed themselves when attempting to eat leatherjackets by spearing their insides with the leatherjacket's sharp dorsal spike.

While most of their time is spent in silent, stealthy ambush, John Dory can make loud barking noises, something that fishers often note when the fish are hauled into their boat. The barks seem to be some sort of territorial display, and snapper will scatter as soon as they hear a John Dory barking.

A John Dory waits silently in the gloom for its prey to swim within striking distance. *(SeacologyNZ)*

TAXONOMY

John Dory (*Zeus faber*) is a cosmopolitan species found in New Zealand waters and many other places around the globe. It belongs to the 'true dory' family Zeidae. Fish in this group all have compressed bodies and extendable jaws.

A GLOBAL FISH

John Dory are found in the North Sea, the Mediterranean, the Arabian Sea, the Indian Ocean and the Pacific. Some fish will evolve into many different species over a small geographic area. Yet, as far as scientists can tell, the same species of John Dory inhabits the waters of the Mediterranean as swims around Aotearoa – although further investigations of their genetics might complicate the picture. This wide distribution means

John Dory have had a global impact, and stories and legends about this fish go back to classical times. It is one of the most widely recognised species in the world, and it is mentioned in children's books, poems, classic literature and sea shanties.

ETYMOLOGY

The scientific name *Zeus* references an ancient Greek name for this fish, *zaeus*, and *faber* refers to an old Dalmatian name for the fish, *fabro*, which means 'craftsman', used because the fish's bones were thought to resemble tools. The Māori name 'kuparu' seems to be unique to Aotearoa.

John Dory is known in Catholic countries as 'St Peter's fish', and in Germany as *Petersfisch*. Several explanations have been offered for the name 'John Dory': one suggestion is that it comes from the French *jaune dorée* meaning 'golden yellow'; another, rather fanciful explanation is that it derives from the word 'janitor', a reference to St Peter's role as the 'doorkeeper' at the gates of heaven. It may also originate from an old English ballad about a French pirate-captain named John Dory, who sets out to capture Englishmen but is caught himself.

Perhaps the most enduring story about the John Dory comes from its association with the apostle Peter, in the gospels of the New Testament. In the biblical story, Jesus is required to pay the temple tax as he travels through the city of Capernaum, so he asks Peter to go fishing in the Sea of Galilee and look inside the mouth of the first fish he catches. Peter dutifully obeys, and sure enough, when he catches a fish and looks inside its mouth, he finds coins to pay the temple tax. The story became associated with John Dory, and the black marks on its sides were believed to be a sign of the miracle, left by Peter's fingers as he plucked the fish out of the sea. Another version of the story says that if these black spots are examined closely the coins can even be seen, in miniature form. As a result of these stories, John Dory became known as 'St Peter's fish', and dried specimens were even hung in churches to keep the miracle alive. However, John Dory don't actually live in the freshwater Sea of Galilee, and the tradition probably originated with a spotted fish called a tilapia.

The John Dory's black spots were believed to have been left by the fingers of the apostle Peter as he plucked the fish out of the waters of the Sea of Galilee. *(Luke Colmer)*

READY FOR THE PAN

While European settlers were wary of most unusual-looking fish, the long history and wide distribution of John Dory meant it was already familiar to them, and it became one of the most popular food fish in New Zealand, fetching high prices at markets. In 1845, English writer Eliza Acton, in her family cookbook, wrote that John Dory was considered by some to be the most delicious fish that appears at the table, although she recommended that 'it can be improved greatly by removing the very ugly head'.[18]

Today many people still enjoy the delicate, sweet taste of John Dory. These fish don't need to be scaled, as their tiny scales are barely noticeable, and they can be

The thin profile of the John Dory makes it difficult to spot when viewed from the front. *(Ian Skipworth)*

cooked directly in the pan like a flounder – barbecued, steamed, baked or grilled. Their flesh makes excellent sashimi and their livers and roe are considered a delicacy.

While John Dory are a popular fishing target for line fishers, one simple strategy is to opportunistically grab them when they swim into shallow water. In winter they move into shallow harbours – possibly to feed on crabs – where their awkward shape makes swimming difficult. Fishers sometimes strand them in the shallows by throwing rocks at them and driving them ashore, and dogs have been known to race into shallow water and grab them. They are not usually targeted directly by commercial fishers, as they tend to be solitary. However, on occasion they will gather together in large numbers; in the 1970s a fishing boat in the Hauraki Gulf reported catching 160 John Dory with one sweep of its net.

TRADED WITH COOK

Unfortunately there is little recorded information about historic John Dory use by Māori. They would have been a familiar fish, especially in the north of the country, and their bones occur sporadically in middens. But it doesn't appear that they were targeted specifically, probably because they spend so much time alone that it wasn't productive to focus on them.

Nonetheless, John Dory were picked up when trawling large seine nets or fishing with hooks and lines. They were traded with the HMS *Endeavour* when the ship passed through Whitianga in 1769, and artist Sydney Parkinson records that they were pickled into casks.[19] The missionary William Colenso travelled through Northland in the 1840s and noted John Dory were popular among Māori children living in the Ngunguru estuary, who caught them in the shallow harbours when the tide receded and the fish could no longer swim upright.

BIOLOGY

John Dory are most commonly found in coastal waters north of Cook Strait. Females are larger than males and can grow to 60 centimetres in length. John Dory are predominantly piscivorous (fish-eaters), scooping up small fish like spotties, flounder, parore, snapper and blue cod in their telescopic mouths. The black marks on the sides of John Dory are 'eye spots' designed to confuse predators. When threatened, the fish turn side on so that the large eye spots make them appear bigger than they really are.

Hāpuku / Groper

Father of whales and tree ferns

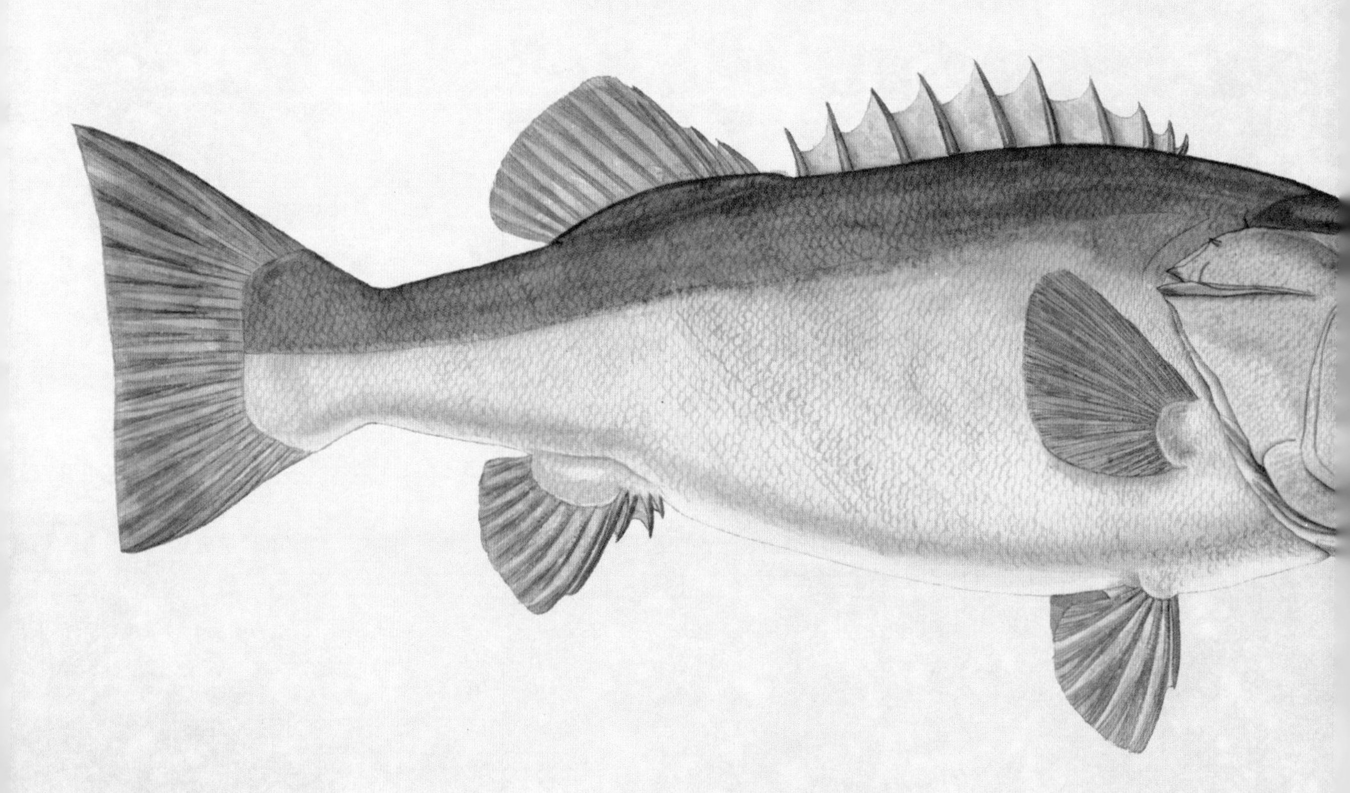

Hāpuku (*Polyprion oxygeneios*) by Frank Edward Clarke, 1876. *(Te Papa)*

The hāpuku is one of the biggest fish in New Zealand seas, with bulging, googly eyes, a big bottom jaw and a powerful, chunky tail. Hāpuku have an incredible appetite, often swallowing their prey whole in one big gulp; they have even been caught with entire shags and penguins in their stomachs. Sometimes they can't quite fit everything in, and crayfish antlers, squid tentacles and barracouta tails can be seen sticking out of their mouths.

For such a voracious carnivore, the hāpuku has a surprisingly naive and curious nature, and will dart up to divers to investigate them. Diver Wade Doak recalled one encounter when a hāpuku swam up to him and sat obediently like a big dog, even allowing another diver to rest their flippers on its back while they adjusted their camera for a better shot.[20]

Hāpuku often bite off more than they can chew, and are sometimes caught with their last meal hanging out of their mouths. *(SeacologyNZ)*

AN UNLUCKY NAME

The hāpuku was one of the most highly sought-after fish species for Māori, and a prestigious catch that brought mana to the fisher. While these fish could be taken from shore, it was believed the largest specimens were to be had far out to sea, when they migrated offshore. The rights to fish at these offshore grounds were passed from generation to generation, and could span centuries.

To reach these grounds, it was essential for the fishing party to leave shore while it was still dark, so there would be enough daylight for the return journey, and a large number of paddlers was needed in case the winds changed while they were out at sea. With such a valuable fish, nothing could be left to chance. Some believed it was unlucky to say the name 'hāpuku' aloud, as the fish might realise they were being hunted, so the fishers would

TAXONOMY

The hāpuku or groper (*Polyprion oxygeneios*) is part of the wreckfish family Polyprionidae, and is found in New Zealand and Australia. It is a close relative of the giant Atlantic wreckfish (*P. americanus*), which is also occasionally found in New Zealand waters.

use the name 'rarawai' instead. Others believed that it was bad luck for hāpuku to touch the side of the boat as they were being hauled in, and a whole fishing trip might be cut short if a hāpuku brushed against the waka accidentally.

But if nothing went wrong, the waka could easily be filled to the brim with hāpuku. When they returned to land, the fishers were greeted on shore by the women of the hapū, who cleaned and scaled the fish and hung them out to dry. All parts of the fish were eaten, either steamed or cooked in a hāngī. Some fish would be kept for winter supplies, cooked and preserved in their own fat and sealed in pockets of bull kelp.

Some iwi returned the heads of the hāpuku to Tangaroa, as thanks for a successful and safe trip, but for others the heads were the best part and considered a delicacy, especially the fat around the eye.

Hooks for catching hāpuku were often tailor-made, using tree roots that were grown into the desired shape and sometimes tipped with bone. *(Auckland War Memorial Museum, Tāmaki Paenga Hira, 14626.1-b)*

In traditional whakapapa, hāpuku is the parent of the tree ferns mamaku and ponga. *(Te Papa)*

THE FATHER OF FERNS

The mana associated with hāpuku is reflected in its prestigious whakapapa. It is said that the fish was the ancestor of many other noble creatures, such as whales, dolphins and tree ferns. One story goes that a long time ago hāpuku and his descendants lived on the land and hadn't yet taken their current form. Hāpuku and his offspring killed a man called Hemā. His son Tāwhaki, seeking revenge, descended from the heavens and attacked the hāpuku and his offspring. The whales, dolphins and sea lions took to the oceans to flee Tāwhaki's wrath, while others such as the tree ferns mamaku, ponga and wheki fled to the bush, where they became the 'fish of the forest'. While tree ferns today bear little resemblance to their ancestor hāpuku, their lineage is still evident on their trunks, where the scars from their fronds look like the scales of a fish.

ETYMOLOGY

The name *Polyprion* means 'many saws' and refers to the fish's sharp fin spines; *oxygeneios* means 'pointed chin', a reference to its protruding lower jaw. The Māori name 'hāpuku' is used in various forms (fāpuku, hāpu'u āpuku) across the Pacific for groper fish in the *Epinephelus* genus. The word 'hāpuku' is also used to mean 'cramming food into the mouth'.

The name 'hapuka' is commonly used today for this fish, but this name appears to have originated with settlers who misheard the Māori name. The English name 'groper' is a corruption of the Portuguese word *garrupa*, used for fish in the Serranidae family. The family name 'wreckfish' refers to the deep-sea habit of these fish, where they can inhabit caves and even shipwrecks.

PRIZED PARTS

The taste of hāpuku had mixed reviews from Pākehā settlers. Some loved the taste of the firm, white meat which was eaten grilled, smoked, salted, fried or boiled and cut into steaks like beef. In the early 1800s, whalers based at remote stations would catch as many as they could in times of plenty and smoke them or pickle them in barrels for winter supplies. The cheeks were especially prized, and commercial fishers targeting hāpuku often ate the cheeks before they could make it to market. In the 1940s, some entrepreneurs started up businesses canning hāpuku meat and extracting oil from the livers, which is rich in vitamins A and D.

But there was also an enduring belief that they were a coarse and unappetising fish, only fit to be eaten by hungry labourers. The surveyor Robert Paulin, while travelling the west coast of the South Island in 1889, commented that hāpuku cutlet 'is not bad by any means when one is hungry or has not tasted fish for some time. They are a coarse fish, and about the cheapest sold in the Dunedin market. I cannot say I care for them myself.'[21] Sometimes the fish could not be sold at market, so it would go to waste or had to be given away. In the early twentieth century, Māori would sometimes wait around at the docks for Pākehā fishermen to return from hāpuku fishing and ask for the eye pieces, to make sure they didn't waste them or throw them away.

Juvenile hāpuku sometimes hide out in floating beds of kelp. *(SeacologyNZ)*

A DRASTIC DECLINE

In the early days of Pākehā settlement, the shoals of hāpuku were mindboggling. A report in the *Otago Witness* in 1890 claimed that 'there have at times appeared immense shoals of the fish at or near the surface of the sea, so that a boat could not be rowed among them without striking them with the oars'.[22] In some cases, when hāpuku were swimming about at the surface, they were simply caught with a harpoon or gaff hook and hauled from the water.

They could once be fished off the rocks around Otago or in the shallow waters of the Hauraki Gulf. But these coastal hāpuku populations suffered a dramatic crash. David Graham of the Portobello Aquarium observed the decline happening very quickly –

they were plentiful around the coast in the 1920s and large 50-kilogram fish were common, but throughout the 1930s the populations began to disappear and large fish became a rarity.[23]

Today hāpuku are typically found only in very deep water. Part of the reason for their decline is that they are long-lived and slow-growing, so their populations don't recover quickly after being overfished. They are also particularly sensitive to fishing as their swim bladders often rupture when they are brought up to the surface, meaning they don't usually survive being caught.

Today, hāpuku is widely admired as one of the most valuable fish in New Zealand, and its meat is sold in top-end restaurants all around the country. New Zealand chef Al Brown describes it as the best-tasting fish anywhere in the world.

To help reduce the impact of fishing on hāpuku populations, scientists at the National Institute of Water and Atmospheric Research (NIWA) have developed a system for farming hāpuku in tanks – a technique not practised for this fish species anywhere else in the world. To date the research has been highly successful: farmed hāpuku flesh is highly regarded internationally and served as sashimi in Japan.

BIOLOGY

The largest hāpuku can grow to more than 1.8 metres in length and weigh up to 100 kilograms. They are slow-growing and can live for over fifty years. Hāpuku move in schools of up to a hundred fish and undertake seasonal migrations offshore. Their big eyes help them hunt in the lower light conditions of deeper waters.

Despite being thought of as deep-water fish, hāpuku were found in coastal waters until overfishing affected their numbers. *(SeacologyNZ)*

DENIZENS OF THE DEEP

Mysterious and elusive

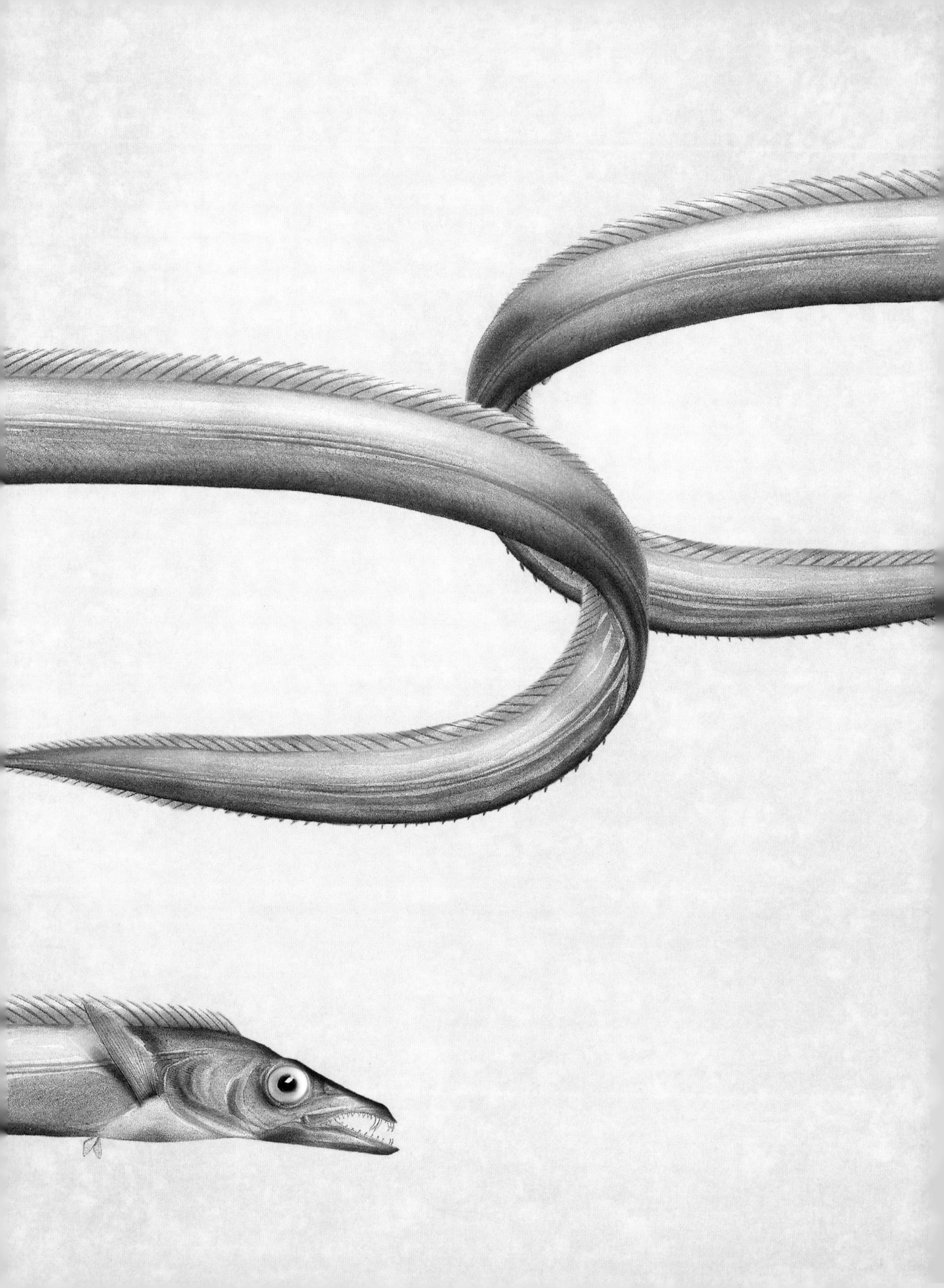

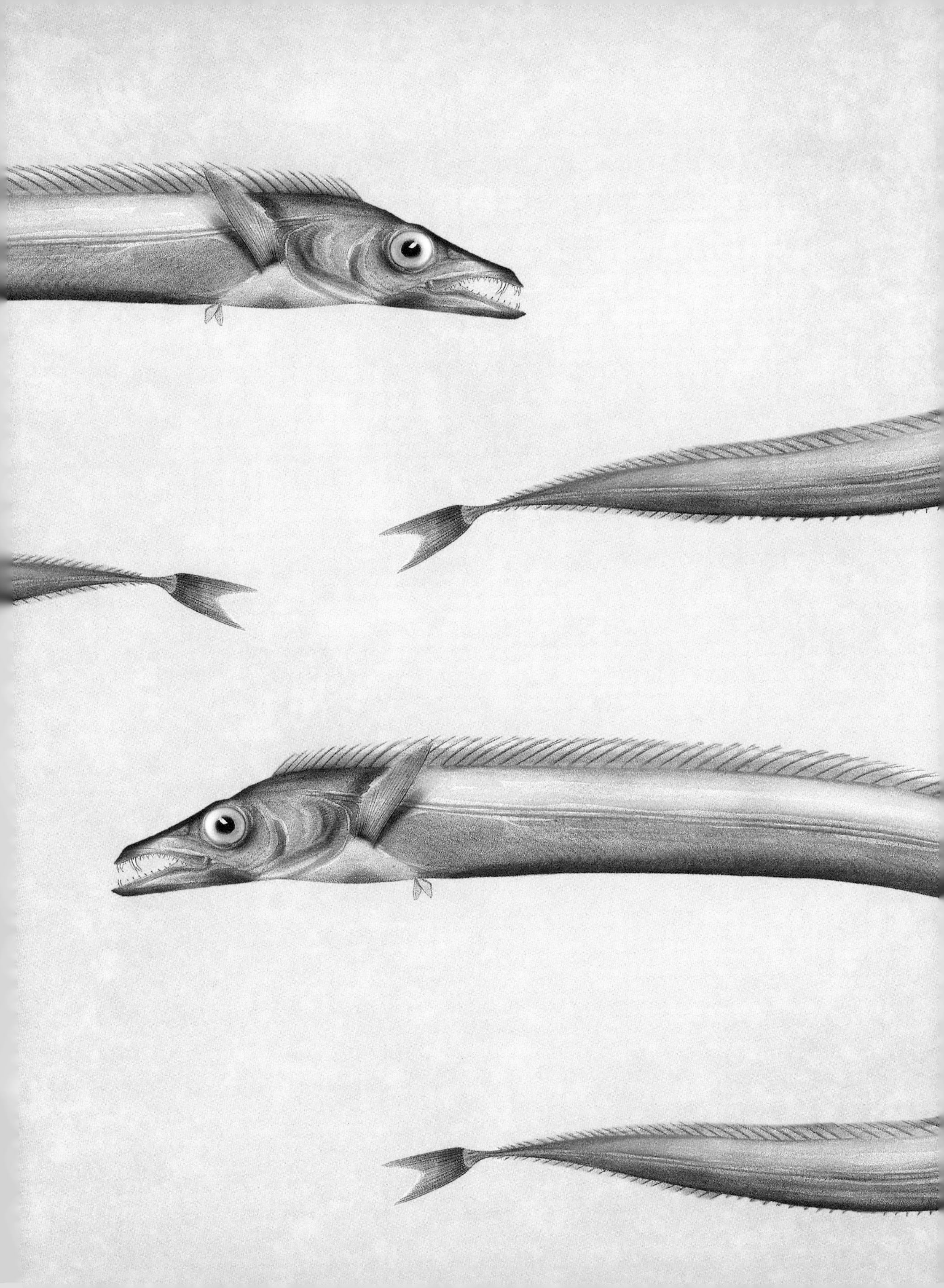

Paper nautilus / Pūpū tarakihi

Sailor on the open sea

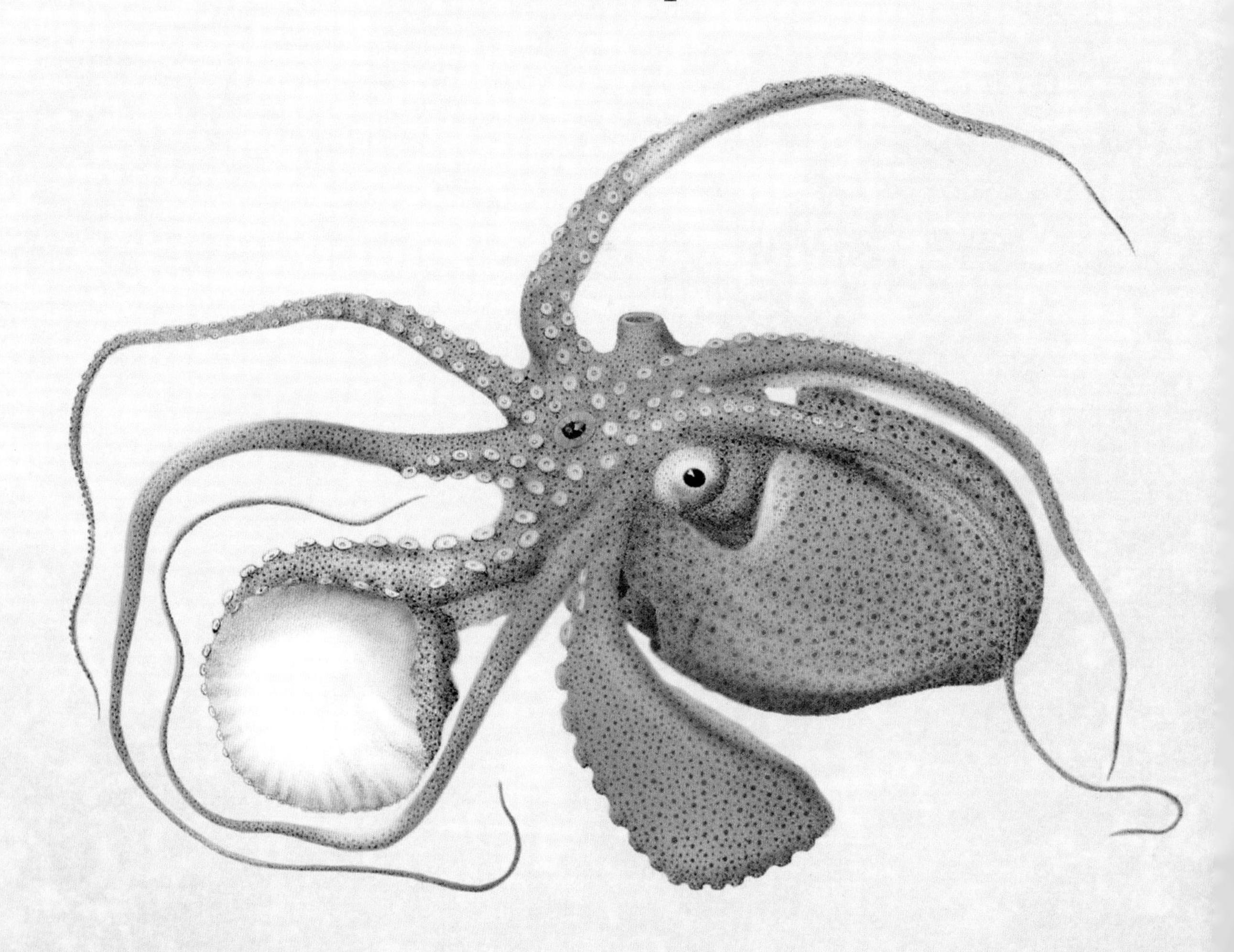

On rare occasions, when the winds are just right, a host of delicate paper-thin shells comes ashore on the New Zealand coast, like a fleet of tiny sails. In reality these are not shells at all, but the egg casings of an open-water octopus known as the paper nautilus. While most octopuses contort themselves into small rocky dens to evade predators and protect their eggs, paper nautiluses swim in the open ocean and have nowhere to hide. But they do have a skill no other animal has: they can employ their tentacles to create beautiful egg casings, which they use like a cradle for storing their eggs. These have a large hole into which the octopus can retreat and hide. However, because they are not a true shell, they can be abandoned if necessary – the octopus simply releases its grip on the egg casing and flees from danger.

These 'shells' also serve another important function. Scientists trying to understand paper nautilus movement have shown that these octopuses use their shell as a home-made swim bladder. When they swim to the surface, paper nautiluses capture air in their egg casings and then seal them with their large tentacles; then, as they jet about the ocean, they can control their buoyancy by releasing air bubbles as needed. This works very well most of the time, although it does occasionally go wrong. When paper nautiluses enter shallow water, they can't dive down deep enough to compress the air in their shells and are forced along by wave action, often ending up stranded on beaches.

TAXONOMY

Two species of paper nautilus occur in New Zealand: the knobby argonaut (*Argonauta nodosus*) is the most common and is found across the southern hemisphere. The greater argonaut (*A. argo*) occurs in tropical and subtropical waters around the world and is occasionally found washed ashore in New Zealand. Both belong to the Argonauta family, the only group of pelagic (open-ocean) octopuses.

HE AHA TE HAU

The strange and unpredictable movements of the paper nautilus – pūpū tarakihi – made a significant impression on Māori. The delicate, papery shells resemble the unfurling of a fern frond and came to symbolise new growth and rebirth. Paper nautiluses feature in legendary tales, like the story of Tinirau looking for his wife Hine-te-iwaiwa. Tinirau found paddling his waka was too slow, so he borrowed the tame paper nautilus of Tautini and sailed it across the ocean to find his wife.

But perhaps the most profound legacy of the paper nautilus was a prophetic vision made by Titahi, a tohunga of Ngāti Whātua. In the eighteenth century, Tītahi saw in a dream that a huge fleet of paper nautiluses would be driven ashore by the north wind, and described his dream in a waiata known as 'He aha te hau'. When the white billowing sails of Pākehā ships were carried by the northern winds into Waitematā Harbour, it was said to be a fulfilment of the prophecy.

Today, the words of 'He aha te hau' are still sung in Tāmaki Makaurau, and the paper nautilus remains an enduring symbol across the Auckland region, incorporated in the design of museums, libraries and even street lights.

The paper nautilus (*Argonauta nodosus*) by Arthur Bartholomew (1881), adapted by Lars Quickfall.

The shell-like egg casing doubles as a protective covering which the octopus can crawl inside. *(SeacologyNZ)*

A SHELL CANOE

New Zealand's early European settlers inherited a longstanding myth about paper nautiluses. The ancient Greeks and Romans believed nautiluses stole their shells from other creatures and used them as boats, while waving their large, wide tentacles in the air as sails and using their other tentacles to row and steer. This idea fascinated British Augustan and Romantic poets, who saw in paper nautiluses a sign of God's immaculate design on Earth. Alexander Pope wrote in his *Essay on Man*, 'Learn of the little nautilus to sail, / Spread the thin oar, and catch the driving gale'[1], and Lord Byron wrote about the freedom of the paper nautilus 'who steers his prow – / The Sea-born Sailor of his shell canoe'.[2] Scientists and further observations later dispelled the myth[3], but it was believed for a long time into the 1800s.

ETYMOLOGY

The name *Argonauta* means 'sailor of the Argo', the legendary ship which Greek hero Jason sailed to retrieve the golden fleece. The word 'nautilus' is derived from an ancient Greek word for sailor. The name *nodosa* refers to the bumpy projections which cover the shell. The Māori name 'pūpū tarakihi' combines the word 'pūpū' – a generic term for shelled molluscs – with 'tarakihi', the name for a species of fish and for the cicada, an insect with a delicate, papery exoskeleton.

An eighteenth-century depiction of a paper nautilus using its webbed tentacles to sail, by Frederick Polydore Nodder, c.1791. *(Smithsonian Libraries)*

European settlers in New Zealand were equally amazed by the beauty and design of the paper nautilus in Aotearoa, and marvelled at how such a hideous and reviled creature as an octopus could produce such a beautiful shell. These 'shells' were highly regarded as ornaments, and could be set in silver and used as tableware such as salt and pepper shakers. Finding large, complete specimens is incredibly rare, and Māori sometimes collected them to sell to settlers as curios, and would obtain high prices.

Mavis Stanton, who lived on Mayor Island in the Bay of Plenty in the 1960s, collected them and packed them into biscuit tins as gifts for friends. She remembers there would occasionally be huge schools of nautiluses driven ashore by the wind and the beach would be white with shells.[4] A Captain Ryder, who skippered a schooner that ran to and from White Island in the 1920s, described another mass stranding on the island in 1929, when thousands of paper nautiluses were blown ashore, dashed against the rocks and broken to pieces. The beach was so covered in shells that you couldn't walk without treading on them, and seagulls repeatedly swooped down, picked up the nautilus shells and dropped them from a height to get to the eggs inside. When Ryder attempted to pick them up, the octopuses would let go of their shells and race back to the ocean, where they would be picked off by schools of kingfish and tarakihi.

European settlers in New Zealand were puzzled as to how such a 'repulsive creature' could produce such a beautiful shell. *(SeacologyNZ)*

ZOMBIE ARMS

While female paper nautiluses have been known for thousands of years, the males were a complete mystery and were only discovered in the last century. Female paper nautiluses were sometimes found covered in strange worm-like creatures, and scientists suspected that these might be the males. The truth turned out to be even stranger, as they are in fact the male's disembodied arms. The males never grow much larger than a peanut, and drift around the ocean without a shell. They have one larger arm that they tuck up in a pouch underneath their eye, and when they discover a female, they extend this arm and attach it to the female, using it to place sperm inside her. After mating, the arm is severed and the male soon dies, but the arm lives on, attached to the female.

Sometimes females are found with numerous male arms attached to them – one was recorded from Flinders Island in Australia with thirty-eight male arms inside her body and wrapped around her gills.[5] The chance encounters between these male and female ocean wanderers are so unlikely that when they do meet a male is likely to sever his arm and attach it to the first female he meets.

BIOLOGY

Paper nautiluses live mostly in tropical and subtropical seas throughout the Indo-Pacific and spend their entire lives floating in the open ocean. They feed on molluscs, small fishes and crustaceans, and in turn are eaten by a range of creatures such as seabirds and fur seals. If their egg casings are damaged or lost they can grow new ones, secreting calcite from their webbed tentacles. While they are thought to be relatively common out at sea, it is rare to see them in the water or for their shells to be washed ashore intact.

Sightings of tiny juvenile paper nautiluses are very rare. *(SeacologyNZ)*

Hagfish / Tuere

Nightmare of the deep

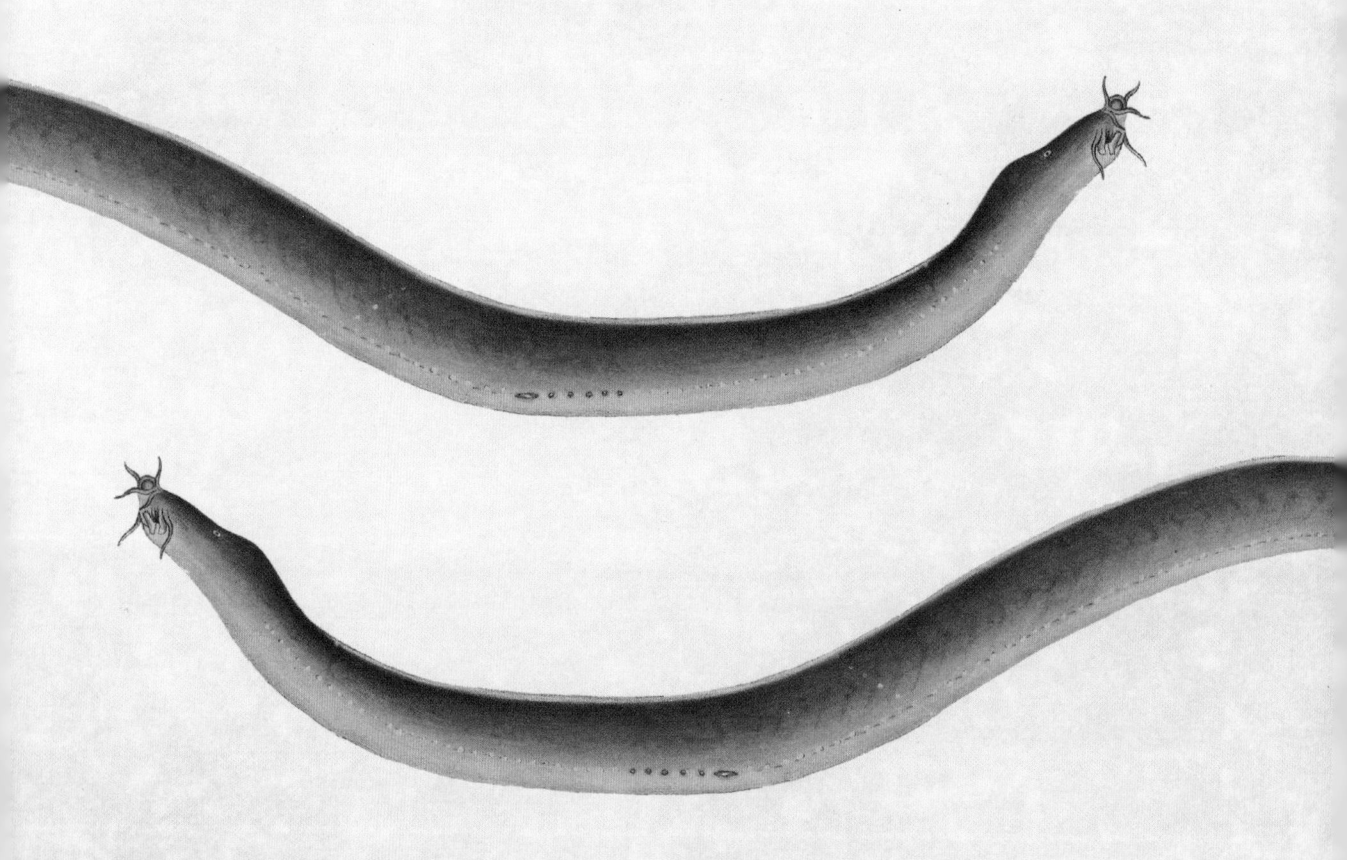

TAXONOMY

There are eight species of hagfish in New Zealand, with the most common being the broad-gilled hagfish (*Eptatretus cirrhatus*). They belong to a group of fish known as agnathans (the jawless fishes), which have many very ancient characteristics that have been lost in other vertebrates, such as a single nostril and several hearts.

Hagfish are blind, eel-like creatures that live at the bottom of the ocean, scavenging for food. When dead animals sink to the depths, a horde of hagfish will swarm the carcass in a writhing mass of wriggling bodies. Once they have burrowed inside their prey, hagfish can feed both with their mouths and by absorbing nutrients through their skin. Sometimes hagfish won't even wait till their prey has died: they can target an injured fish, slip through its gills or rasp through its flesh and feed on it from the inside out.

Lacking a true backbone, hagfish don't have a lot of leverage when it comes to forcing their way into a fish carcass. So they tie themselves in a knot and slam themselves into their prey. This knot-tying ability comes in handy when they are searching holes and burrows looking for food, as they can tie their back end into a knot to hold them in place and pull themselves out again.

The blind hagfish uses its feelers to seek out food on the sea floor. *(Matt Ruglys)*

Broad-gilled hagfish (*Eptatretus cirrhatus*) by Theodoor van Lith der jeude, c.1860s. *(University of Amsterdam).*

SLIME ATTACK

Although hagfish are slow-moving and blind, few fish species prey on them, as they have an ingenious strategy for defending themselves: they ooze thick slime from their pores, which rapidly expands 10,000-fold into a cloud of sticky jelly. If a predatory fish tries to attack, it quickly finds its gills covered in slime and begins choking to death. Hagfish can choke on their own slime as well, so they need to move away quickly after it has been released.

Once away from danger, hagfish rid themselves of excess slime by tying themselves in a knot and scraping their skin clean. This slime release is also employed for attacking and subduing live prey, and hagfish in New Zealand waters have been observed feeling out their prey in burrows, sliming them to death and then hauling them out to eat.

ETYMOLOGY

The scientific name *Eptatretus* means 'seven holes' and refers to the gill apertures. The name *cirrhatus* means 'tendrils' and describes the barbels at the front of the mouth. The origin of the Māori name 'tuere' is not clear, but another name for this fish is 'pia', the word used for sticky tree-resins like kauri gum.

The fish has a number of English names, such as 'blind eel', 'slime eel' and 'snot eel'. Fishers would sometimes refer to them as 'true sailors', because of their ability to twist nets and fishing lines into slime-covered knots that were impossible to untie. The name 'hagfish' refers to old witches who were believed to have the ability to turn water into glue.

A TREASURED MEAL

Despite the hagfish's disgusting habits, Māori often viewed them as the cleaners of the sea – tidying up the ocean of fish carcasses and other dead animals. In a number of places around the country they were considered to be a delicacy, and one of the most palatable fish in the sea. Hagfish were rinsed in fresh water until all their slime had come out, turned inside out, and then dried and cooked. The slime itself was also considered good eating, and some believed that it was a nutritious meal that could make you grow strong.

Hagfish and their slime are regarded as a taonga by some iwi and considered delicious. *(Sarah Milicich)*

To catch hagfish, Māori would tie bits of rotten meat to a rope and dangle it in an area of the sea with strong currents.[6] One of the best times to collect hagfish was after a whale stranding, as the huge carcass on the beach would attract them close to shore.

After the arrival of Europeans, when the whaling industry was in full swing, it was said that hagfish were so abundant they used to come into the shallows in their millions.[7] As workers flensed whales on the beach, hagfish would try to dig into the flesh and tear off pieces. Māori workers would kill these hagfish quickly with a blubber spade and add them to a pile to be taken home for dinner. Hagfish have remained a delicacy in some areas, and are regarded by some iwi as a significant taonga, to be served at feasts and on special occasions.

A FEARFUL CREATURE

Māori sometimes persuaded more daring Europeans to indulge in eating hagfish, including the explorers Charles Heaphy and Thomas Brunner, who ate it with crayfish and baked mamaku fern during their South Island journey in the 1840s. But for most European settlers, the hagfish was one of the ugliest and most horrific creatures in the sea, and some referred to it as 'the terror of fishermen' or 'the nightmare of the deep'.[8]

Fishers especially despised them, and in some places where hagfish were prolific, lines had to be checked constantly to make sure the catch had not been attacked. One area off Otago Heads became known to fishers as 'The Hospital', as it would only ever yield injured fish, missing body parts or internal organs. Hagfish are sometimes caught in crayfish pots as well, with fishers hoping for a lucrative haul of crayfish finding twenty or more hagfish oozing slime over the boat and twisting themselves into knots. Many fishers will cut their lines rather than haul hagfish into the boat, as their sticky slime dries hard like glue, ruining fishing equipment and damaging boats.[9]

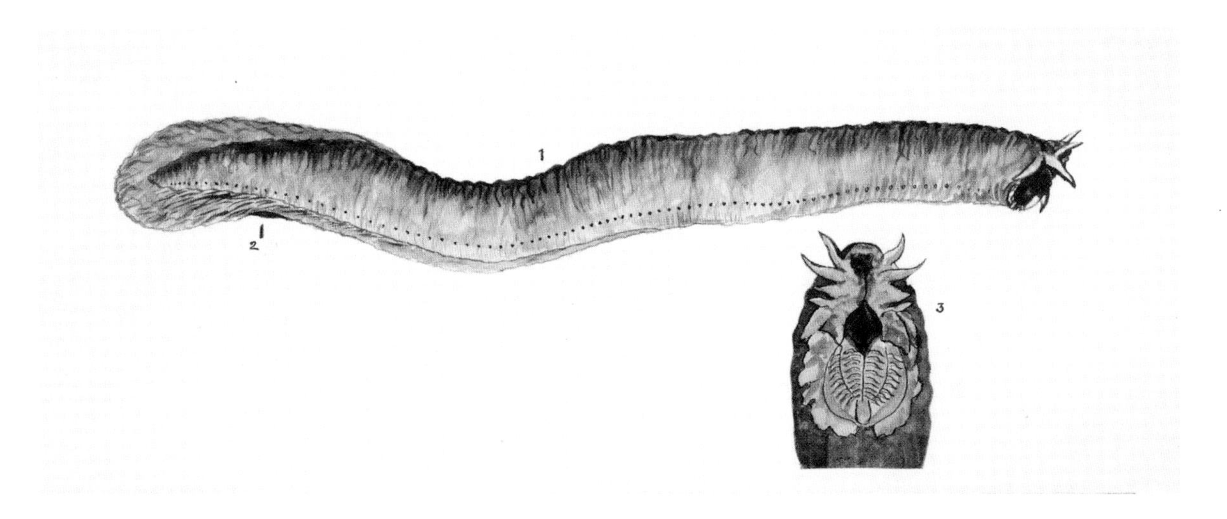

Broad-gilled hagfish *(Eptatretus cirrhatus)* by L. T. Griffin. *(Auckland War Memorial Museum, Tāmaki Paenga Hira, PD-1971-5-23)*

CLOTHING OF THE FUTURE

For such a strange fish that lives at the bottom of the sea, the hagfish has a number of surprising uses. Overseas, hagfish skins are deslimed, tanned into leather and sold as 'eel skin', which is made into purses, wallets and boots. Hagfish slime has attracted a huge amount of interest among scientists, with some claiming that it may prove to be the clothing of the future. Embedded in the slime are tiny threads, hundreds of times thinner than a human hair but ten times stronger than nylon, with similar properties to spider silk. This has led researchers to attempt to mimic these compounds to make durable clothing that doesn't rely on petroleum-based products. Research is currently under way to find out if synthetic hagfish slime could be used to produce everything from bulletproof vests to fabric, bandages, food packaging and bungy cords. The slime has even been investigated by the US navy as a way to gum up the propellers of enemy ships.

While it is not commonly eaten in New Zealand, hagfish is popular in Korea, where it is served skinned, grilled or stir-fried in spicy sauce, and is believed by some to have aphrodisiac properties. The slime can be used as an alternative to egg whites in cooking, and has even been used to make scones.

BIOLOGY

The common hagfish is dark brown to purple-brown, and found from the north of New Zealand to Rēkohu/Chatham Islands. Hagfish are most commonly found in water deeper than 90 metres, but will come into the shallows when there is an abundance of food such as discarded fish and guts from fishing vessels. In some places they can form a large part of the diet of seals and dolphins, which, as air-breathing mammals, are less affected by their slime.

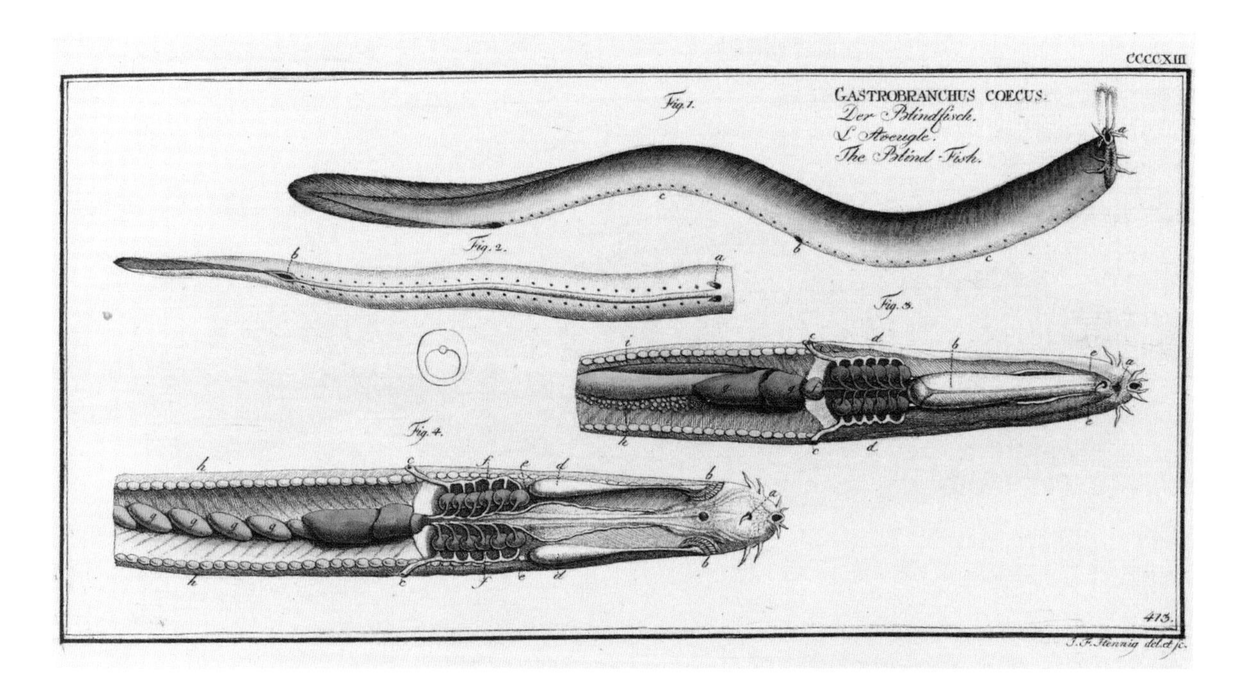

Until recently there hasn't been a lot of investigation into hagfish in New Zealand, so we don't know a lot about their life history, ecology or population sizes. But work overseas has shown hagfish can occur at enormously high densities in deep-sea sediments, and may be the single most abundant fish species that occurs there. This means that hagfish likely play a critical role in deep-sea ecosystems, both as predators hunting small fish and invertebrates and as scavengers helping to recycle dead animal matter. They may play a role in soft sea sediments like that of earthworms on land – burrowing through the sand and turning it over.

However, hagfish have a slow rate of growth, and a number of fisheries overseas have collapsed. While there is a small hagfish fishery in New Zealand, little is known about the impact of fishing on their populations and more work is needed to make sure that they are not overharvested.

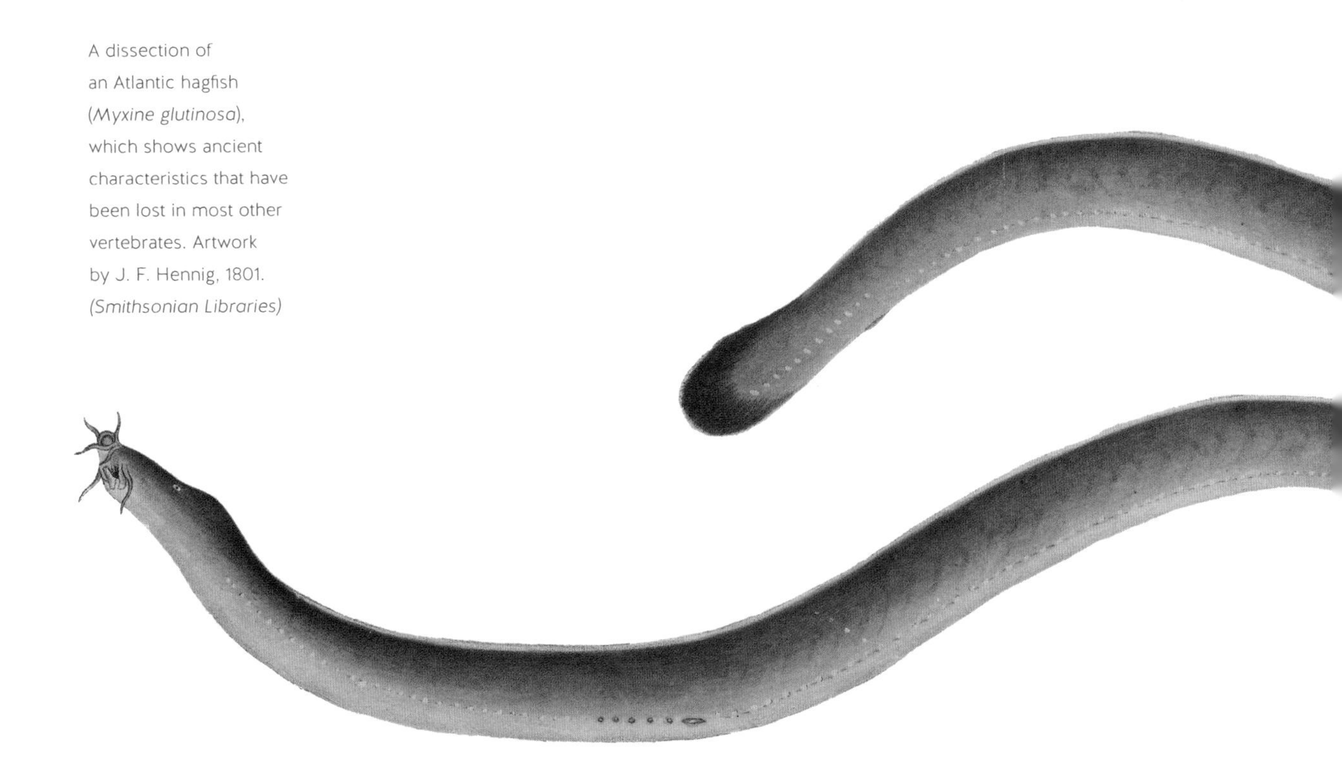

A dissection of an Atlantic hagfish (*Myxine glutinosa*), which shows ancient characteristics that have been lost in most other vertebrates. Artwork by J. F. Hennig, 1801. *(Smithsonian Libraries)*

Frostfish / Pāra

The fish that casts itself ashore

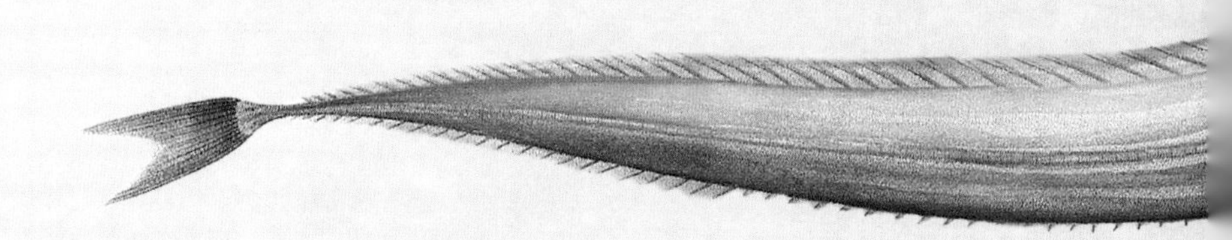

Frostfish (*Lepidopus caudatus*) by Jacques Reyne Isidore Acarie-Baron (c.1880s), adapted by Lars Quickfall. *(University of Amsterdam)*

Frostfish are a beautiful sight to behold: thin streaks of silver that are so shiny you can see your reflection in their skin. Although they can wiggle their bodies back and forth to move, they are not designed for speed. Instead, they spend a lot of their time hanging vertically in the water column like a floating silver sword, using their gently beating fins to keep them upright. With their thin bodies and reflective skin, they effectively become invisible, allowing them to sneak up on fish and snap them up with their dagger-like teeth.

TAXONOMY

The frostfish or pāra (*Lepidopus caudatus*) belongs to the cutlassfish family, Trichiuridae. All members of this group are long, eel-like predatory fish, coloured blue or silver and resembling a steel sword.

These secretive fish are not commonly caught, but they occasionally wash ashore on beaches, which is where they are most often seen. There was a belief among Pākehā settlers that frostfish chose to end their own lives and threw themselves ashore in 'deliberate acts of self-immolation'.[10] Some believed that they ended their lives in pairs, and if one was found on the beach you only needed to wait and its mate would soon come ashore as well. Lighthouse keeper C. H. Robson recalled how he once saw a frostfish swimming to shore and attempted to turn it around, but each time the fish ended up swimming back to land. He wrote: 'As he seemed to have set his mind upon landing, I gave up the attempt to influence his decision, and took him home for breakfast.'[11]

The enduring mystery of frostfish strandings has tantalised scientists and fishers for decades, and a wide range of theories have been proposed to explain it. One said they ran themselves aground as they were chasing their prey; another said they were infested with so many parasites they were driven crazy; yet another claimed the males were chasing females in an attempt to breed and beached themselves by mistake. Many commentators believed it was a snap frost that killed the fish or drove them senseless and washed them ashore – it was noted that they always seemed to come ashore on frosty mornings – but this was probably a case of confirmation bias, as they are washed ashore on sunny days as well.

Today, scientists believe the most likely explanation has to do with their spawning. During the winter months, frostfish migrate inshore to spawn but their awkward shape isn't intended for fast swimming, and once they have spent their energy spawning, they are often caught up in currents and washed onto beaches.

ETYMOLOGY

The scientific name *Lepidopus* means 'scale-foot', which relates to the scale-like spine on the fish's pelvic fins; *caudatus* means 'tailed', a reference to its distinct forked tail. Māori call frostfish 'pāra', 'taharangi' and 'hiku'. Hiku is another word for 'tail' and taharangi can also mean 'lacking energy' or 'undecided'. The name 'frostfish' is said to come from the belief that the fish washes up after frosty mornings, but it also may derive from its silvery coating, which rubs off when touched.

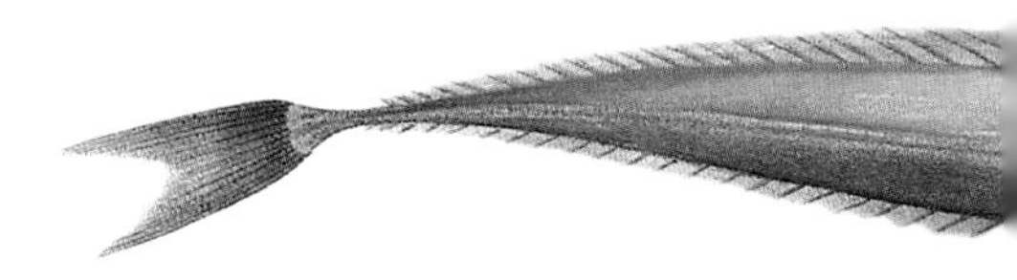

A frostfish, also known as scabbard fish, by Robert Hamilton, 1843. *(Smithsonian Libraries)*

CHASING THE MOON

Māori had their own traditional explanation for frostfish stranding. It was believed that the fish beached themselves as they were trying to chase after the moon, or that they were cast ashore as punishment for being so difficult to catch on a fishing line.

Frostfish had an important place in the origin story of many fish species. Originally, frostfish and other fish were believed to have lived in celestial waters – puna-kauariki. However, a severe drought dried up these waters and the fish feared they would be attacked by matuku, the Australasian bittern. So they descended from the heavens and went to live in the oceans together. All was well for a time, until the greedy frostfish started eating the eels' children. The other fish fled in terror from the ravenous frostfish – eels found refuge in the swamps and freshwaters, lamprey hid under boulders, whitebait found safety in numbers, and hagfish left for the depths of the sea.

When frostfish did strand, they were a useful supplementary food. The beaches of Moeraki in North Otago were a common area for frostfish to run aground, and after the arrival of European settlers Māori would often scour the beach on horseback looking for them. To carry their haul of frostfish they would hook the tail through the gills and wrap it around the head of the horse like a necklace. These catches were stored in baskets and sold to travellers passing through the area by train.

In some areas there was a tradition of removing the eyes and throwing them back into the sea in the hope that this would attract more frostfish – but this practice fell out of favour because a fully intact frostfish would make more money at market.

GOING FROSTFISHING

Europeans adored the taste of frostfish, and fisheries expert R. A. A. Sherrin believed they were the most delicious fish in New Zealand. Politician Edward Gibbon Wakefield tried frostfish in 1890 at a restaurant in Dunedin and was instantly hooked, describing the experience vividly: '[The waiter] placed before each of us a dish containing what looked to me like veal cutlets, but what proved to be little slices of the richest, and at the same time the most delicate, fish I had ever tasted. The flesh was quite white and semi-transparent, and it almost melted in the mouth, it was so tender.' He was so delighted by the taste of the fish that some locals invited him on a frostfishing trip. He dutifully turned up early the next morning equipped with a fishing rod, fly-fishing lures and a range of hooks, but was handed a shotgun instead. The locals took him down to Blueskin Bay, north of Dunedin, and showed him how to 'catch' the fish off the beach, by collecting the stranded fish by hand and using the shotgun to blast away the seagulls that reached the fish before they did.[12]

BIOLOGY

Frostfish normally grow to around 1.5 metres in New Zealand waters. They live all around New Zealand, but usually in water more than 200 metres deep. They often form schools that travel towards the surface at night to feed on crustaceans, small fish and squid. Despite being found around the world, frostfish are a difficult species to study and little is known about them.

Two men survey their catch of frostfish at Waihī Beach in 1914. *(Auckland Libraries Heritage Collection, AWNS-19140702-50-1)*

Frostfishing could be highly profitable, and some keen frostfishermen would camp out in the dunes so that they could be out at daybreak before the gulls and other fishers arrived. One of the largest reported 'catches' of frostfish, of 160 fish, occurred on the Kapiti Coast in the 1870s. In 1886 a man claimed to be earning £6 a week selling frostfish – a small fortune at the time, if true. Fisherman Ray Doogue recalls a time in the mid-twentieth century when frostfish could be sold for around $2 apiece in Dunedin, at a time when big crayfish would fetch only ten cents.[13] In later times, cars were sometimes used for scouting the beaches, and a number were lost to the sand and tide in their owners' eagerness for a haul of frostfish.

Sunfish / Rātāhuihui

Sunbathing giants

A sunfish is an enormous pancake of a fish that looks like it has been cut in half and is missing its back end. It's the largest bony fish in the sea, growing several metres long and weighing over 2 tonnes. To reach this tremendous size, the sunfish undergoes one of the most incredible growth spurts of any animal. Unlike other large sea creatures like sharks and whales, which are already quite big when they are born, sunfish start life as one of 300 million tiny eggs.[14] When they hatch, the larvae resemble tiny pufferfish only 2 millimetres long. So, to reach their adult size, they need to grow up to sixty million times larger, while somehow avoiding being eaten at every stage of the process.

For a long time, people thought sunfish were simply giant plankton that drifted about on ocean currents, as they were often seen lying placidly on the surface of the water. However, they are actually quite active hunters, regularly making deep dives by flapping their fins like a penguin to catch squid and jellyfish with their tiny, beak-like mouths. After making these dives they can become cold and sluggish, and will sunbathe on the sea surface to warm up.

These slow-moving giants grow to be the largest bony fish in the sea. *(iStock)*

Ocean sunfish (*Mola mola*) by Kawahara Keiga, c.1820s. *(Naturalis Biodiversity Center)*

CLEANING SERVICE

These slow-moving giants are easy targets for parasites. Over seventy different parasite species have been recorded living on sunfish to date, leading one writer to describe the species as a 'travelling zoological garden'.[15] Removing these parasites is a difficult task for the sunfish, and they are sometimes seen leaping out of the water in an attempt to dislodge them. They also seek out floating bits of kelp, where they recruit the services of cleaner fish, which nibble off parasites from their skin, eyes, gills and the inside of their mouths. Sometimes sunfish will swim after seabirds and roll to one side, like a dog looking for a belly rub. Those birds that take up the offer will roost on top of the sunfish and peck the numerous parasites off their skin.

Sunfish bathe at the surface to warm up after deep dives and sometimes to be pecked clean of parasites. *(iStock)*

TAXONOMY

The three most common species of sunfish in New Zealand waters are the giant sunfish (*Mola alexandrini*), which occurs mostly in the north; the hoodwinker sunfish (*M. tecta*), which occurs mainly in the south; and the ocean sunfish (*M. mola*), which occasionally visits New Zealand waters from the Pacific. They all belong to the sunfish family, Molidae, and to the order Tetraodontiformes, which includes leatherjackets and porcupine fish.

CHANCE ENCOUNTERS

There is little recorded information about the historical significance of sunfish to Māori, although there is some evidence of their capture in the early twentieth century. It is likely that sunfish were encountered during trips to offshore fishing grounds, as they can be quite conspicuous lying on the surface of the water. One possible interpretation of the Māori name 'rātāhuihui' is 'to gather at the sun', which suggests an understanding of sunfish ecology.

Sunfish strandings are relatively common in some places, such as along the Whangārei coast, so Māori would have come into contact with them this way as well. Sunfish are said to taste like stingray, which was considered one of the best-tasting seafoods, so they may have been eaten if they were washed ashore in good condition.

TROPHY HUNTING

Pākehā had no interest in eating sunfish, which were thought to be worthless as food. But there was something about their large size and strange shape that encouraged some people to hunt them as a trophy. There are stories of people attacking them with anything they could find at hand: boat oars, rocks, even makeshift harpoons fashioned out of pocket-knives and wooden sticks. In Gisborne in 1892, some locals even dropped sticks of dynamite on a sunfish that swam in close to the breakwater, then hauled it out of the sea with chains.[16]

Often these hunters had no real use for the fish after killing it, and had to tow the carcass out to sea so it didn't rot and cause a stink. Dead sunfish were sometimes hauled into town and exhibited as a freak of nature, with locals charged a shilling or sixpence to see the monster from the sea. Whanganui Regional Museum curator Samuel Drew heard about such a fish and travelled to Napier to purchase it. But by the time he got there it was a week old, and by his account just about everyone in the neighbourhood had cut a memento out of it. The liver had been taken to make oil and a woman had even removed the fin to use as a fire screen.[17]

ETYMOLOGY

The name *Mola* is a Latin word meaning 'millstone', and refers to the fish's strange circular shape. The Latin word tecta means 'hidden', a reference to the fact that the fish avoided detection for so long. The meaning of the Māori name 'rātāhuihui' isn't exactly clear, and it is shared with the fin whale (*Balaenoptera physalus*). One suggested meaning is 'sea monster', and another is 'to gather at the sun', a possible reference to the fish's sunbathing habit. The European name 'sunfish' also refers to this habit.

The sunfish has a variety of different names across the world: Hawaiians call it *kunehi makua* ('king of the mackerels'), in Norway it is *klumpfisk* ('lump fish'), and in Germany it is *Schwimmender Kopf* ('swimming head').

Sunfish responded to being attacked by humans in different ways. In some instances, they didn't fight back at all, allowing themselves to be stabbed and speared, seemingly unfazed. However, perhaps these fish were still too cold or dazed from deep dives to muster up the energy to swim away, as they can be vigorous fighters. In 1944 a group from the Piha Surf Club was rowing a small boat when they spotted a sunfish and attacked it with their oars. The sunfish turned and charged the boat several times and when one man managed to hook the fish in the mouth with the anchor the sunfish pulled him overboard.[18]

Bump-head sunfish (*Mola alexandrini*) swimming in deep water. *(Camille Hay)*

In another story, from 1934, a group of fishermen in Pelorus Sound hooked a sunfish with a makeshift harpoon. The sunfish raced away and was able to pull their 7-ton boat around for several miles. The crew stabbed at the fish and shot at it with a .303 rifle, which barely made an impact on its thick, leathery skin. Eventually it managed to pull free and swam away.[19]

This trophy mentality lasted into the twentieth century, but has gradually faded away. Most people would now consider a chance sighting of sunfish in their natural habitat to be one of the most beautiful and remarkable experiences in nature.

THE HOODWINKER SUNFISH

In 2014, a sunfish washed up on a Christchurch beach, and researchers realised it was an entirely new species – the first discovered in over 120 years. They named it the hoodwinker sunfish (*Mola tecta*) for its success in having evaded detection for so long.

The immense size of the sunfish and its similarity to other sunfish species were partly what had allowed it to remain hidden. In order to confidently describe a sunfish as a new species, scientists had to compare preserved specimens collected in museums. But sunfish weigh several tonnes and need to be soaked in vast quantities of formaldehyde and alcohol to preserve them. Many old museum specimens have shrunk or shrivelled over time, or their taxidermists mounted them in strange shapes, making it difficult to compare specimens.

BIOLOGY

The ocean sunfishes (genus *Mola*) are found worldwide – and contain the largest bony fish in the ocean. One of the heaviest specimens ever found, a *Mola alexandrini* washed ashore at Whangārei Heads, was 3.3 metres long, 3.2 metres high, and weighed over 2 tonnes. As juveniles, sunfish have a diverse diet but as they get bigger they move further into the ocean and prefer to eat jellyfish. Once they reach adulthood, there are few creatures large enough to attack them – mostly sharks, killer whales and sea lions. Unfortunately, their preference for jellyfish potentially places them at risk of consuming plastic bags by mistake.

Once scientists learned how to tell the difference between sunfish species more accurately, they realised the hoodwinker sunfish had been hiding in plain sight. It was recently discovered that a cast of a sunfish that had been hanging in the entrance hall of the Otago Museum for decades was a hoodwinker sunfish.[20] What's more, the hoodwinker is now believed to be one of the most common sunfish in New Zealand waters. It is incredible to reflect that if one of the largest fish in the ocean could have escaped discovery for so long, what other unknown creatures might be lurking beneath the waves?

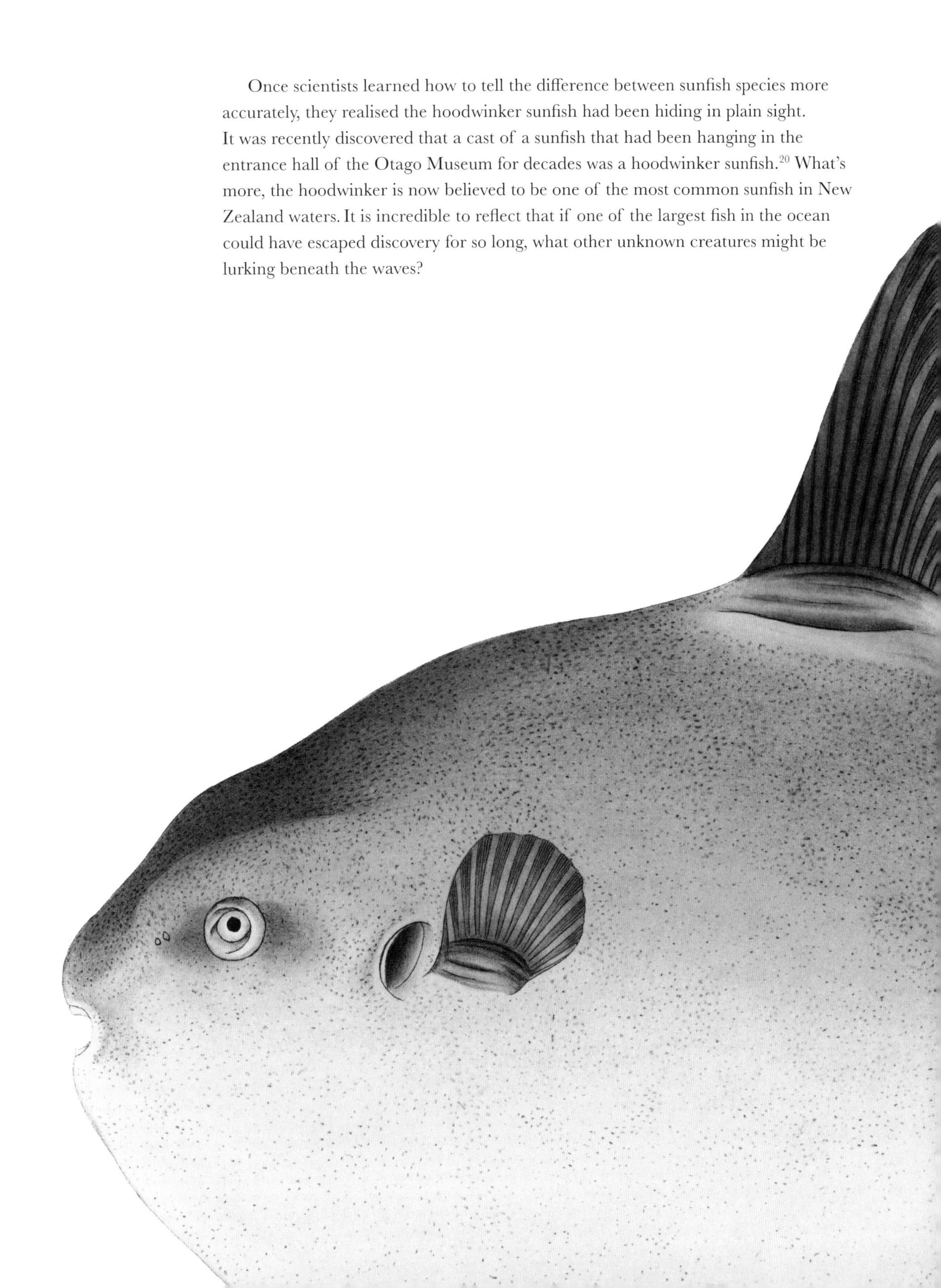

ENDNOTES

INTRODUCTION

1. Banks, J. (1768–71). *The* Endeavour *journal of Sir Joseph Banks.* Project Gutenberg Australia, https://gutenberg.net.au/ebooks05/0501141h.html
2. Beaglehole, J. C., ed. (1961). *The journals of Captain James Cook on his voyages of discovery.* Volume 2. Cambridge: Hakluyt Society, 169.
3. Banks, J. *The* Endeavour *journal.*
4. It should be noted that not all of these interactions were so positive. European explorers didn't realise that they were upsetting a delicate balance and breaking strict protocols around fishing that had maintained these resources for centuries. In 1772 French explorer Marc-Joseph Marion du Fresne stayed in the Bay of Islands for over a month, believing he had found paradise on earth. But he was killed, along with many of his crew, possibly because they had broken fishing restrictions and other protocols in the area. With some notable and tragic exceptions, Cook's first voyage was able to establish friendly relations with local Māori, in large part due to the presence of the Ra'iātean high chief Tupaia.
5. Taylor, N. M. (1959). *Early travellers in New Zealand.* Oxford: Clarendon Press. Retrieved from: Early New Zealand Books: University of Auckland.
6. The culture of fish. (1874). *Otago Witness*, Issue 1175, 6 June 1874, 22.
7. Hursthouse, C. F. (1857). *New Zealand: Or, Zealandia, the Britain of the south* (Volume 2). London: E. Stanford. Retrieved from: Early New Zealand Books: University of Auckland.
8. Sherrin, R. A. A. (1886). *Handbook of the fishes of New Zealand (Vol. 2).* Auckland: Wilson and Horton.
9. Waitangi Tribunal (1992). Wai 27 – *The Ngai Tahu sea fisheries report.* Waitangi Tribunal Report, published online 2012.
10. Goat Island is a beloved national icon today, but its establishment was controversial. Getting the reserve gazetted was a bitter and drawn-out fight, and the scientists who proposed the idea were bombarded with hate mail. After the reserve was established, a local newspaper even ran the headline 'Nothing to do at Goat Island any more'.

FRESH WATER

1. Mueller, G. in: McDowall, B. (2011). *Essays of a fishery scientist: 50 years of experience.* Don Jellyman (ed.), NIWA Information Series, 80.
2. Hobson, J. in: Pond, W. (1997). *The land with all woods and water.* Waitangi Tribunal Report, published online 1997.
3. N. Chalmers quoted in Beattie, H. (1920). *Nature-lore of the southern Maori.* New Zealand Institute. *Transactions and Proceedings of the Royal Society of New Zealand*, Volume 52, 1920, 53. Retrieved from: https://paperspast.natlib.govt.nz/periodicals/TPRSNZ1920-52.2.9.1.13
4. Anon. (1933). The lamprey. *Otago Daily Times*, Issue 22114, 18 November 1933, 9.
5. Taylor, R. (1855). *Te Ika a Maui: Or, New Zealand and its inhabitants, illustrating the origin, manners, customs, mythology religion, rites, songs, proverbs, fables, and language of the natives.* Cambridge University Press. Retrieved from: Early New Zealand Books: University of Auckland.

6. These by-products of metabolism are deposited as the pigment *Biliverdin*, which turns their skin blue. This pigment is also responsible for the greenish colour of bruises in human skin.

7. St John, J. H. A. (1873). *Pakeha rambles through Maori lands*. Robert Burrett, Molesworth Street. Retrieved from: Early New Zealand Books: University of Auckland.

8. The term 'whitebait' can also encompass the juveniles of native smelt (*Retropinna retropinna*), and in former times would have included the extinct grayling or upokororo (*Prototroctes oxyrhynchus*), but these were typically regarded as 'second-class' whitebait and were not the primary goal of most fishers.

9. Brunner, T. (1850). Journal of an expedition to explore the interior of the Middle Island of New Zealand. *The Journal of the Royal Geographical Society of London*, 20, 344–378.

10. Anon. (1888). Flood and field [from correspondent about grayling]. *Lyttelton Times*, Volume LXX, Issue 8605, 5 October 1888, 2.

11. Lee, F., & Perry, G. L. (2019). Assessing the role of off-take and source–sink dynamics in the extinction of the amphidromous New Zealand grayling (*Prototroctes oxyrhynchus*). *Freshwater Biology*, 64(10), 1747–1754.

SANDY SHORES

1. South Island pied oyster catcher (*Haematopus finschi*), from Baker, unpublished data in: Owen, K. L., & Sell, M. G. (1985). The birds of Waimea inlet. *Notornis*, 32, 271–309.

2. Graham, D. (1974). *A treasury of New Zealand fishes*. A. H. & A. W. Reed.

3. Samuel, E. (1936). The toheroa: New Zealand's exclusive shell-fish. *The New Zealand Railways Magazine*, Volume 11, Issue 4.

4. A nourishing toheroa soup was sometimes fed to patients during the 1918 influenza pandemic, with one patient describing it as their salvation.

5. The Prince at Wellington. (1920). *Manawatu Standard*, Volume XLIII, Issue 1809, 8 May 1920, 5.

6. Some of the excuses by poachers have been equally inventive: one man argued that he had to collect toheroa because his pregnant wife had cravings for them.

7. Another important flounder habitat was Napier's now-lost inner harbour, which was once known as Te Pātiki Tahanui o Te Whangaui-o-Rotu – 'the big-sided flounder of the great bay of Rotu'.

8. Waitangi Tribunal (1992). Wai 27 – *The Ngai Tahu sea fisheries report*. Waitangi Tribunal Report, published online 2012.

9. Stingray barb spears have a long history of use in both Polynesian and western traditions. In Greek mythology, the Ithacan King Odysseus was killed accidentally by his son, who speared him with a stingray barb. In the eighteenth century, the Ra'iātean high priest Tupaia, who would later

travel to New Zealand with James Cook, had a stingray barb thrust through his chest by invaders from Bora Bora, and narrowly escaped death. Stingray spears do not, however, appear to have been a very common weapon in New Zealand, probably due to the difficulties of acquiring enough stingray barbs to produce them.

10. Duignan, P. J., Hunter, J. E., Visser, I. N., Jones, G. W., & Nutman, A. (2000). Stingray spines: A potential cause of killer whale mortality in New Zealand. *Aquatic Mammals*, 26(2), 143–147.

11. One marvellous fishy tale was reported in the *New Zealand Herald* in 1936, and shows just how voracious their appetite can be. If you can believe the story, an oarsman at the New Plymouth rowing club dropped his set of false teeth in the sea during practice. Several months later, a fisherman hauled up a large gurnard and when preparing the fish for dinner found the teeth almost undamaged, and was able to return them to the shocked owner. (Strange restoration. [1936]. *New Zealand Herald*, Volume LXXIII, Issue 22332, 1 February 1936, 12.)

12. In other traditions they have a more prestigious whakapapa and are the siblings of snapper and trevally – valued food fishes of northern New Zealand.

ROCKY REEFS

1. One rather fantastic account from George Grey describes Māori making a giant diving cage – taiki – which a diver entered and in which he or she was lowered to the sea floor in order to collect crayfish. No other accounts of the technique exist and it does seem rather hard to believe, given crayfish could just as easily be scooped out of the water at low tide.

2. Banks, J. (1768–71). *The* Endeavour *journal of Sir Joseph Banks.* Project Gutenberg Australia, https://gutenberg.net.au/ebooks05/0501141h.html.

3. Stack, J. W. and E. (1938). *Further Maoriland adventures of J. W. and E. Stack, Book I: James West Stack's story*, Chapter IX, 58–63. A. H. & A. W. Reed, Dunedin, http://www.enzb.auckland.ac.nz

4. Heaphy, C. (1846). Notes of an expedition to Kawatiri and Araura, on the western coast of the Middle Island. *Nelson Examiner and New Zealand Chronicle*, Volume V, Issue V, 12 September 1846, 111.

5. The exploits of a group of commercial kina divers is the focus of the documentary series *Spiky Gold Hunters*, which has developed an international fanbase.

6. Te Wheke o Muturangi continued to have a powerful influence in the area after being slain. The Brothers Islands / Ngawhatu Kai-ponu in Cook Strait were said to be the eyes of Te Wheke, and were believed to be incredibly dangerous to look at directly.

7. This tradition has a long history, as octopuses were also seen as a prestigious and chiefly food in Hawaii and other Pacific islands.

8. Hugo, V. (1866). *Toilers of the sea*. Project Gutenberg Australia, https://www.gutenberg.org/ebooks/32338

9. The visit of H.M.S. New Zealand. (1913). *Otago Daily Times*, Issue 15750, 29 April 1913, 4.

10. A huge octopus. (1925). *Te Aroha News*, Volume XLI, Issue 6616, 22 June 1925, 5.

11. Graham, D. H. (1930). Marine Nature Notes: The octopus called 'devil fish' or 'sucker fish'. *Otago Daily Times*, Issue 21124, 6 September 1930, 10.

12. Dynamiting fish. (1912). *Auckland Star*, Volume XLIII, Issue 121, 21 May 1912, 7.

13. Morrison, M., Lowe, M. L., Jones, E. G., Makey, L., Shankar, U., Usmar, N. R., ... & Middleton, C. (2014). *Habitats of particular significance for fisheries management: the Kaipara Harbour*. New Zealand Aquatic Environment and Biodiversity Report No. 129. Ministry for Primary Industries.

14. H. A. (1905). The compleat angler. *The Sydney Morning Herald*, 11 March 1905.

15. Hector, J. (1872). *Fishes of New Zealand: Notes on the edible fishes.* James Hughes.

16. Beaglehole, J. C., ed. (1961). *The journals of Captain James Cook on his voyages of discovery*. Volume 2. Cambridge: Hakluyt Society, 169.

17. Papers relating to special settlement on Stewart's Island. *Appendix to the Journals of the House of Representatives*, 1872 Session I, D-07a. Accessed: https://atojs.natlib.govt.nz

18. One archaeologist estimated the amount of snapper a person would have eaten over the course of a year was about forty-six, and they would have been larger and heavier than the average fish today. (Leach, F. [2006]. Fishing in pre-european New Zealand. *Archaeofauna*, 15, 19–276.)

19. Best, E. (1908). Forest lore. *Transactions of the New Zealand Institute*, Volume 42.

20. Cook, J. (1893). *Captain Cook's journal during his first voyage round the world, made in HM Bark Endeavour, 1768–71*. Elliot Stock, London. Retrieved from: https://gutenberg.net.au/ebooks/e00043.html

21. Beaglehole, J. C., ed. (1961). *The journals of Captain James Cook on his voyages of discovery*. Volume 2. Cambridge: Hakluyt Society, 29.

22. Barlow, P. W. (1888). *Kaipara: Or, experiences of a settler in north New Zealand.* S. Low, Marston, Searle, & Rivington.

23. 'The Australasian'. (1868). Sea fisheries of New Zealand. *North Otago Times,* Volume X, Issue 294, 17 March 1868, 4.

24. Travers, W. T. S. (1864). Lecture on acclimatization. *Lyttelton Times,* Volume XXI, Issue 1216, 31 March 1864, 5.

25. Author interview with Colin Brickell, October 2021.

26. An initiative in Auckland, Kai Ika, offers a fish-filleting service for a small fee, then unwanted fish heads and offcuts are distributed by Papatūānuku Kōkiri Marae to appreciative members of the community.

27. Colman, J. A. (1972). Food of snapper, *Chrysophrys auratus (forster),* in the Hauraki Gulf, New Zealand. *New Zealand Journal of Marine and Freshwater Research,* 6(3), 221–239.

OCEAN HUNTERS

1. In another tradition underlying the significance of sharks, the Milky Way galaxy was thought of as a giant shark known as Mangōroa, placed there by the demigod Māui.

2. Noted in the caption for the famous 1844 lithograph of the event by Joseph Jenner Merrett, *Maori feast at Remuera, 1844*. Retrieved from: https://collections.tepapa.govt.nz/object/620928

3. Swainson, W. (1859). *New Zealand and its colonization*. London: Smith, Elder.

4. Matthews, R. H. (1910). *Reminiscences of Maori life fifty years ago*. New Zealand Institute. One waka alone was said to have collected over 260 sharks, including huge tiger sharks, which dragged them around before they were hauled on board.

5. French explorer Louis Duperrey frequently tried to bargain for sharks' teeth during his travels in New Zealand in the early 1800s, and found Māori were very reluctant to part with them no matter what he offered.

6. The politician Edward Tregear reported communities at Kawhia and Whanganui that admired shark teeth so much that several men had filed their teeth to a point to resemble them. (Tregear, E. [1916]. Maori mummies. *The Journal of the Polynesian Society*, Volume 25, No. 100, 167–168.)

7. This tenacity for life is demonstrated by a 140-kilogram mako shark caught by American angler Alfred Glassell. It was hit twice by a whale spade, almost severing its head, and towed for two hours to the dock, before being hoisted for weighing. On the cable it came back to life, sending everyone scattering for cover, then it managed to wriggle itself back into the water and swim off.

8. Grey, Z. (2014). *Tales of the anglers' Eldorado*. New York: Harper & Brothers. Available from Project Gutenberg Australia, https://gutenberg.net.au/ebooks06/0608281.txt

9. Anon. (1927). A river of fish. *Hawera Star*, Volume XLVI, 17 January 1927, 9.

10. Roughley, T. C. (1916). *Fishes of Australia and their technology*. (No. 21). W. A. Gullick, Government Printer.

11. Another caught near Raoul Island had three little shearwaters in its stomach. While this phenomenon is relatively rare today, New Zealand once had vast colonies of seabirds all around its coast that could have provided kingfish with an abundant, easily obtained food source. (Duffy, C. A., & Taylor, G. A. [2015]. Predation on seabirds by large teleost fishes in northern New Zealand. *Bulletin of the Auckland Museum*, 20, 497–500.)

12. R. A. A. Sherrin makes the claim that they were regularly left stranded on the beaches after storms, and collected by Māori this way as well. (Sherrin, R. A. A. [1886]. *Handbook of the fishes of New Zealand (Vol. 2)*. Wilson and Horton.)

13. There was a similar prejudice against kingfish in Australia, where for a time it had almost no market value at all. However, much of the stigma seems to have been unrelated to the taste. When a pair of New Zealand fish sellers made a big haul of kingfish in New South Wales, they sold them under the name 'emperor fish' to all the best fish shops in town, and everyone who tried them thought them delicious.

14. Another fascinating interspecies behaviour was recorded by legendary diver Wade Doak, who saw a group of kingfish surround a pod of sunfish and repeatedly graze their bodies against them. Presumably they were using the rough, sandpapery skin of the sunfish as an exfoliant to rid themselves of parasites.

15. Skinner, H. D. (1919). Some Maori fish-hooks from Otago. *Transactions and proceedings of the Royal Society of New Zealand*, 51, 267.

16. Mackenzie, J. (1885). *Papers relating to the development of New Zealand colonial industries*. I. Fisheries Correspondence. No. 1. Mr. J. Mackenzie to the Hon. Sir Julius Vogel.

17. Graham, D. (1974). *A treasury of New Zealand fishes*. A. H. & A. W. Reed.

18. Acton, E. (1845). *Modern cookery for private families*. Longmans: London.

19. Parkinson, S. (1984). *A journal of a voyage to the South Seas*. Caliban Books. Retrieved from: Early New Zealand Books: University of Auckland.

20. Doak, W. (1972). *Fishes of the New Zealand region*. Hodder & Stoughton, Auckland.

21. Paulin, R. (1889). *The wild west coast of New Zealand: A summer cruise in the* 'Rosa'. Thorburn & Company, London.

22. The Fisheries Court. (1890). *Otago Witness*, 23 January 1890, 17.

23. Graham, D. H. (1974). *A treasury of New Zealand fishes.* A. H. & A. W. Reed.

DENIZENS OF THE DEEP

1. Pope, A. (1829). *An essay on man, Epistle III*, line 177, in: *An essay on man: And other poems*, Volume 2. John Sharpe, Duke Street, Piccadilly.

2. Byron, G. G. (1837). *The island, Canto 1* in: *The complete works of Lord Byron: Reprinted from the last London edition.* A. and W. Galignani Co: Paris.

3. The idea was eventually disproved by scientist Jeanne Villepreux-Power, who did tests on paper nautiluses in the 1820s. In doing so, she became the first person to create the modern aquarium.

4. Mikaere, B. (1983). The obsidian island. *New Zealand Geographic*, 03. Retrieved from: www.nzgeo.com/stories/the-obsidian-island/

5. Caiger, P. (2018). The legend of the Argonaut. *New Zealand Geographic*, 150. Retrieved from: www.nzgeo.com/stories/the-legend-of-the-argonaut/

6. After European settlement, carcasses of sheep and other farm animals could be used as bait.

7. An exaggeration for sure, but hagfish can certainly occur at remarkable densities. In the Gulf of Maine, Martini et al. (1997) estimated that the Atlantic hagfish (*Myxine glutinosa*) could reach densities of up to 500,000 fish per square kilometre.

8. The southern wonderland. (1896). *Otago Witness*, Issue 2183, 2 January, 45.

9. Hagfish slime is so sticky that when, in the US state of Oregon in 2017, a truck carrying hagfish crashed, the slime released over the highway glued cars to the ground.

10. Mysterious visitors. *The Press*, Volume LIV, Issue 16286, 9 August 1918, 5.

11. Thompson, G. M. (1920). Wild life in New Zealand. *Otago Witness*, Issue 3473, 5 October 1920, 53.

12. Wakefield, E. (1890). Catching frost fish with a shotgun. *Outing*, 15, 308–311. Retrieved from: https://archive.org/details/outing15newy/page/n3/mode/1up

13. Doogue, R. B. (1967). *Hook, line and sinker*. A. H. & A. W. Reed, 243.

14. Up to 300 million eggs are laid at one time by a single female ocean sunfish (*Mola mola*), making this species the most fertile back-boned animal on the planet.

15. The sun fish. (1896). *Otago Witness*, Issue 2206, 11 June 1896, 3 (supplement).

16. A sea monster. (1930). *Gisborne Times*, Volume LXX, Issue 11182, 15 April 1930, 5.

17. It took five men three full days to remove the skin of the sunfish, and Drew described the experience as 'a most unpleasant task … [the skin was] a hard gristly substance that very quickly turned the edges of the sharpest knives, blistering our hands that had already been made sore by the cutting roughness of the skin'. He could only continue with the work by making frequent trips out of the room to be sick. There were so many pieces cut out of the specimen that Drew could re-create only one half of the fish – filling in the gaps with pieces from the other side. The skin was soaked in formaldehyde and hung on the wall of the Whanganui Regional Museum. (From the Whanganui Regional Museum website: http://collection.wrm.org.nz/search.do?id=1206&db=object&page=1&view=detail)

18. Fight with sunfish. (1944). *Manawatu Standard*, Volume LXIV, Issue 30, 4 January 1944, 4.

19. A true fish story of Pelorus Sounds. (1934). *Nelson Evening Mail*, Volume LXVI, 5 July 1934, 9.

20. Rykers, E. (2018). Hoodwinked. *New Zealand Geographic*, Issue 49. Retrieved from: www.nzgeo.com/stories/hoodwinked/

BIBLIOGRAPHY

A gigantic sunfish. (1908). *Wanganui Chronicle*, Vol. L, Issue 12145, December 1908, 8.

A huge octopus. (1925). *Te Aroha News*, Vol. XLI, Issue 6616, 22 June 1925, 5.

A river of fish. (1927). *Hawera Star*, Vol. XLVI, 17 January 1927, 9.

A sea monster. (1930). *Gisborne Times*, Vol. LXX, Issue 11182, 15 April 1930, 5.

A true fish story of Pelorus Sounds. (1930). *Nelson Evening Mail*, Vol. LXVI, 5 July 1934, 9.

A voracious fish. (1922). *New Zealand Herald*, Vol. LIX, Issue 18243, 9 November 1922, 9.

Abe, T., & Sekiguchi, K. (2012). Why does the ocean sunfish bask? *Communicative & Integrative Biology*, 5(4), 395–398.

Abe, T., Sekiguchi, K., Onishi, H., Muramatsu, K., & Kamito, T. (2012). Observations on a school of ocean sunfish and evidence for a symbiotic cleaning association with albatrosses. *Marine Biology*, 159(5), 1173–1176.

Acton, E. (1845). *Modern cookery for private families*. Longmans: England.

Adkins, N. F. (1938). 'The chambered nautilus': Its scientific and poetic backgrounds. *American Literature*, 9(4), 458–465.

Anderson, A. (1981). Barracouta fishing in prehistoric and early historic New Zealand. *Journal de la Société des Océanistes*, 37(72), 145–158.

Anderson, A., Binney, J., & Harris, A. (2015). *Tangata whenua: A history*. Bridget Williams Books.

Anderson, T. J. (1999). Morphology and biology of *Octopus maorum* Hutton 1880 in northern New Zealand. *Bulletin of Marine Science*, 65(3), 657–676.

Andrew, N., & Francis, M. (eds). (2003). *The living reef: The ecology of New Zealand's rocky reefs*. Craig Potton Publishing.

Andrew, N. L. (1988). Ecological aspects of the common sea urchin, *Evechinus chloroticus*, in northern New Zealand: A review. *New Zealand Journal of Marine and Freshwater Research*, 22(3), 415–426.

Andrews, J. R. H. (1989). *The southern ark: Zoological discovery in New Zealand, 1769–1900*. Century Hutchinson New Zealand.

Apte, S., Smith, P. J., & Wallis, G. P. (2007). Mitochondrial phylogeography of New Zealand freshwater crayfishes, *Paranephrops* spp. *Molecular Ecology*, 16(9), 1897–1908.

Ashe, J. L., & Wilson, A. B. (2020). Navigating the southern seas with small fins: Genetic connectivity of seahorses (*Hippocampus abdominalis*) across the Tasman Sea. *Journal of Biogeography*, 47(1), 207–219.

Ayling, A. M. (1981). The role of biological disturbance in temperate subtidal encrusting communities. *Ecology*, 62(3), 830–847.

Babirat, C., Mouritsen, K. N., & Poulin, R. (2004). Equal partnership: Two trematode species, not one, manipulate the burrowing behaviour of the New Zealand cockle, *Austrovenus stutchburyi*. *Journal of Helminthology*, 78(3), 195–199.

Baker, A. N. (1971). Food and feeding of kahawai (Teleostei: Arripididae). *New Zealand Journal of Marine and Freshwater Research*, 5(2), 291–299.

Baker, C. F., Jellyman, D. J., Reeve, K., Crow, S., Stewart, M., Buchinger, T., & Li, W. (2017). First observations of spawning nests in the pouched lamprey (*Geotria australis*). *Canadian Journal of Fisheries and Aquatic Sciences*, 74(10), 1603–1611.

Banks, J., *The* Endeavour *journal of Sir Joseph Banks*. Project Gutenberg Australia, https://gutenberg.net.au/ebooks05/0501141h.html

Barber, I. (2004). Sea, land and fish: Spatial relationships and the archaeology of South Island Maori fishing. *World Archaeology*, 35(3), 434–448.

Barlow, P. W. (1888). *Kaipara: Or, experiences of a settler in north New Zealand*. S. Low, Marston, Searle, & Rivington.

Bartle, J. A. (1993). Differences between British and French organization of zoological exploration in the Pacific 1793–1840. *Tuatara*, 32, 75–81.

Bathgate, A. (1874). *Colonial experiences: Or, sketches of people and places in the province of Otago, New Zealand*. J. Macelehose. Retrieved from: Early New Zealand Books, University of Auckland.

Beattie, H. (1920). *Nature-lore of the southern Maori*. New Zealand Institute.

Best, E. (1902). Notes on the art of war, as conducted by the Maori of New Zealand, with accounts of various customs, rites, superstitions, &c., pertaining to war, as practised and believed in by the ancient Maori. Part III. *The Journal of the Polynesian Society*, Vol. 11, 3(43), 127–162.

——. (1904). Maori medical lore. Notes on sickness and disease among the Maori people of New Zealand, and their treatment of the sick; together with some account of various beliefs, superstitions and rites pertaining to sickness, and the treatment thereof, as collected from the Tuhoe tribe. Part I. *The Journal of the Polynesian Society*, Vol. 13, 4(52) (December 1904), 213–237.

——. (1924). *The Maori. Vol. 2*. Board of Maori Ethnological Research for the Author and on behalf of the Polynesian Society.

——. (1929). *Te Whare Kohanga ('The Nest House') and its lore*. A. R. Shearer, Government Printer.

——. (1977a). *Maori religion and mythology*. AMS Press.

——. (1977b). *Forest lore of the Maori* (No. 14). Te Papa Press.

——. (1979). *Fishing methods and devices of the Maori*. Dominion Museum Bulletin No. 12. Government Printer.

——. (1982). *Maori religion and mythology 2*. Government Printer.

Blue cod. (1908). *Marlborough Express*, Vol. XLII, Issue 35, 12 February 1908, 7.

Bradshaw, J. (2001). *The far downers*. University of Otago Press.

Bradstock, M. (1989). *Between the tides: New Zealand shore and estuary life*. David Bateman.

Broadhurst, M., Krueck, N., & Roelofs, A. (2021). *Status of Australian fish stocks report – Luderick (2020)*. Fisheries Research and Development Cooperation.

Brunner, T. (1850). Journal of an expedition to explore the interior of the Middle Island of New Zealand. *The Journal of the Royal Geographical Society of London*, 20, 344–378.

Buck, P. (Te Rangi Hiroa). (1921). Maori food-supplies of Lake Rotorua, with methods of obtaining them, and usages and customs appertaining thereto. *Transactions and Proceedings of the Royal Society of New Zealand*, Vol. 53, 1921, 433.

——. (Te Rangi Hiroa). (1926). The Maori craft of netting. *Transactions and Proceedings of the New Zealand Institute*, Vol. 56, 597–646.

——. (Te Rangi Hiroa). (1949). *The coming of the Maori*. Whitcombe & Tombs: Wellington.

Byron, G. G. (1837). *The complete works of Lord Byron*. A. and W. Galignani Co: Paris.

Caiger, P. (2018). The legend of the Argonaut. *New Zealand Geographic*, 150. Retrieved from: https://www.nzgeo.com.

Campbell, M., & Nims, R. (2019). Small screens, small fish and the diversity of pre-European Māori fish catches. *Journal of Pacific Archaeology*, 10(2), 43–54.

Carson, S. F., & Morris, R. (2017). *Collins field guide to the New Zealand seashore*. HarperCollins.

Cassels, R. (1972). Prehistoric man and his environment. In Goodall, D. H. (ed.), *The Waikato, man and his environment*. Waikato Branch, New Zealand Geological Society.

Challis, A. J. (1995). *Ka pakihi whakatekateka o Waitaha: The archaeology of Canterbury in Maori times*. Department of Conservation.

Chatham Islands. (1892). *Lyttelton Times*, Vol. LXXVII, Issue 9681, 23 March 1892, 6.

Clapcott, J., Ataria, J., Hepburn, C., Hikuroa, D., Jackson, A. M., Kirikiri, R., & Williams, E. (2018). Mātauranga Māori: Shaping marine and freshwater futures. *New Zealand Journal of Marine and Freshwater Research*, 52, 457–466.

Colenso, W. (1844). *Excursion in the northern island of New Zealand in the summer of 1841–2*. Launceston Examiner. Retrieved from: Early New Zealand Books, University of Auckland.

——. (1879). Contribution to a better knowledge of the Maori race. *Transactions of the New Zealand Institute*, Vol. 12, 108–147.

——. (1880). On the vegetable food of the ancient New Zealanders before Cook's visit. *Transactions and Proceedings of the Royal Society of New Zealand*, Vol. 13, 1880, unnumbered page.

——. (1881). On the fine perception of colours possessed by the ancient Maoris. *Transactions and Proceedings of the Royal Society of New Zealand*, Vol. 14, 49.

——. (1889). *Ancient tide-lore and tales of the sea, from the two ends of the world*. R.C. Harding.

Colman, J. A. (1972). Food of snapper, *Chrysophrys auratus* (forster), in the Hauraki Gulf, New Zealand. *New Zealand Journal of Marine and Freshwater Research*, 6(3), 221–239.

Connor, C. A. (2010). A diachronic exploration of the harvesting of the marine environment to a distinctive New Zealand English lexicon, 1796–2005. [Unpublished doctoral dissertation.] Victoria University of Wellington.

Cornwall, C. E., Phillips, N. E., & McNaught, D. C. (2009). Feeding preferences of the abalone *Haliotis iris* in relation to macroalgal species, attachment, accessibility and water movement. *Journal of Shellfish Research*, 28(3), 589–597.

Cowan, J. (1930). *The Maori, yesterday and to-day*. Whitcombe & Tombs.

——. (1931). Pictures of New Zealand life. *The New Zealand Railways Magazine*, Vol. 6, Issue 6.

Crowe, A. (2018). *Pathway of the birds: The voyaging achievements of Māori and their Polynesian ancestors*. University of Hawai'i Press.

Dieffenbach, E. (1843). *Travels in New Zealand (Vol. 1)*. Retrieved from: Early New Zealand Books, University of Auckland.

Disease of goitre. (1933). *Otago Daily Times*, Issue 21880, 16 February 1933, 12.

Doak, W. (1971). *Beneath New Zealand seas*. A. H. & A. W. Reed.

——. (1972). *Fishes of the New Zealand region*. Hodder & Stoughton, Auckland.

Doogue, R. B. (1967). *Hook, line and sinker*. A. H. & A. W. Reed.

Doogue, R. B., & Moreland, J. M. (1960). *New Zealand sea anglers' guide*. A. H. & A. W. Reed.

Drummond, J. (1911). [Frost fish.] The naturalist: Notes on natural history in New Zealand. *Otago Witness*, Issue 3009, 15 November 1911, 76.

——. (1911). In touch with nature. *Lyttelton Times*, Vol. CXXII, Issue 15723, 16 September 1911, 12.

——. (1911). The naturalist: Notes on natural history in New Zealand. *Otago Witness*, Issue 3015, 27 December 1911, 76.

——. (1912). [Paper nautilus.] The naturalist: Notes on natural history in New Zealand. *Otago Witness*, Issue 3025, 6 March 1912, 76.

——. (1923). In touch with nature: The grayling. *Otago Witness*, Issue 3640, 18 December 1923, 6.

——. (1930). Nature notes: The paper nautilus. *New Zealand Herald*, Vol. LXVII, Issue 20490, 15 February 1930, 1 (supplement).

Duffy, C. A., & Taylor, G. A. (2015). Predation on seabirds by large teleost fishes in northern New Zealand. *Bulletin of the Auckland Museum*, 20, 497–500.

Duignan, P. J., Hunter, J. E., Visser, I. N., Jones, G. W., & Nutman, A. (2000). Stingray spines: A potential cause of killer whale mortality in New Zealand. *Aquatic Mammals*, 26(2), 143–147.

Dumont d'Urville, J. S-C. (1950). *New Zealand 1826–1827: From the French of Dumont d'Urville. An English translation*. Retrieved from: Early New Zealand Books, University of Auckland.

Dunedin to Melbourne. (1911). *Otago Witness*, Issue 1911, 25 September 1890, 34.

Dunn, M. R., Connell, A. M., Forman, J., Stevens, D. W., & Horn, P. L. (2010). Diet of two large sympatric teleosts, the ling (*Genypterus blacodes*) and hake (*Merluccius australis*). *PLoS One*, 5(10), e13647.

Dynamiting fish. (1894). *Thames Star*, Vol. XXV, Issue 4845, 18 September 1894, 2.

Dynamiting fish. (1912). *Auckland Star*, Vol. XLIII, Issue 121, 21 May 1912, 7.

Eels prey on trout: 'A slow growing beast'. (1938). *Poverty Bay Herald*, Vol. LXV, Issue 19585, 17 March 1938, 4.

Far northern cannery sends food to troops overseas. *Northern Advocate*, 21 June 1944, 5.

Fifty years ago. [Gander Flat fifty years ago.] (1945). *Bay of Plenty Times*, Vol. LXXIII, Issue 13775, 7 June 1945, 4.

Fight with sunfish. (1944). *Manawatu Standard*, Vol. LXIV, Issue 30, 4 January 1944, 4.

Finn, J. K. (2013). Taxonomy and biology of the argonauts (Cephalopoda: Argonautidae) with particular reference to Australian material. *Molluscan Research*, 33(3), 143–222.

——. (2018). Recognising variability in the shells of argonauts (Cephalopoda: Argonautidae): The key to resolving the taxonomy of the family. *Memoirs of Museum Victoria*, 77.

Finn, J. K., & Norman, M. D. (2010). The argonaut shell: Gas-mediated buoyancy control in a pelagic octopus. *Proceedings of the Royal Society B: Biological Sciences*, 277(1696), 2967–2971.

Fish poaching. (1932). *The Press*, Vol. LXVIII, Issue 20728, 13 December 1932, 4.

Fiso, M. (2020). *Hiakai: Modern Maori cuisine*. Random House New Zealand.

Flood and field. [From correspondent about grayling.] (1888). *Lyttelton Times*, Vol. LXX, Issue 8605, 5 October 1888, 2.

Flood, A. S., Goeritz, M. L., & Radford, C. A. (2019). Sound production and associated behaviours in the New Zealand paddle crab *Ovalipes catharus. Marine Biology*, 166(12), 1–14.

Freshwater crayfish. (1936). *Otago Daily Times*, Issue 22916, 25 June 1936, 10.

Fyfe, R., & Bradshaw, J. (2020). A review of the role of diadromous ikawai (freshwater fish) in the Māori economy and culture of Te Wai Pounamu (South Island), Aotearoa New Zealand. *Records of the Canterbury Museum*, 34, 35–55.

Giant sunfish. (1932). *Bay of Plenty Times*, Vol. LX, Issue 10966, 16 November 1932, 3.

Girl's heart transfixed by stingray. (1938). *Northern Advocate*, 1 December 1938, 8.

Goddard, M. (2011). *Rangikapiti Pa historic reserve.* Historic Heritage Assessment. Kaitaia Area Office, DOC.

Gold bangle. (1935). *Auckland Star*, Vol. LXVI, Issue 57, 8 March 1935, 8.

Goldie, W. H. (1998). *Maori medical lore: Notes on the causes of disease and treatment of the sick among the Maori people of New Zealand, as believed and practised in former times, together with some account of various ancient rites connected with the same.* Southern Reprints.

Good with garlic and spices. (1964). *Te Ao Hou*, March 1964, 6.

Gordon, Keith (2003). 'Frog ladies and hubble bubbles'. In Sue Thompson (ed.), *50 Years of New Zealand underwater*, New Zealand Underwater Association, 26–27.

Graham, D. H. (1930). [Butterfish.] Marine nature notes. *Otago Daily Times*, Issue 21112, 23 August 1930, 3.

——. (1930). Marine nature notes: Echinoidea. *Otago Daily Times*, Issue 21160, 18 October 1930, 3.

——. (1930). Marine nature notes: Pawa (Paua) or mutton fish shell. *Otago Daily Times*, Issue 21184, 15 November 1930, 2.

——. (1930). The octopus called 'devil fish' or 'sucker fish'. *Otago Daily Times*, Issue 21124, 6 September 1930, 10.

——. (1931). Marine nature notes: Colour changes in flat fish. *Otago Daily Times*, Issue 21468, 17 October 1931, 2.

——. (1931). Marine nature notes. [Lamprey.] *Otago Daily Times*, Issue 21456, 3 October 1931, 2.

——. (1931). Marine nature notes: Blue cod. *Otago Daily Times*, Issue 21504, 28 November 1931, 18.

——. (1931). Marine nature notes: Garfish. *Otago Daily Times*, Issue 21527, 26 December 1931, 5.

——. (1931). New Zealand whitebait. *Otago Daily Times*, Issue 21384, 11 July 1931, 2.

——. (1932). Marine nature notes: Paper nautilus. *Otago Daily Times*, Issue 21800, 12 November 1932, 5.

——. (1932). Marine nature notes: Barracoutta. *Otago Daily Times*, Issue 21776, 15 October 1932, 2.

——. (1932). Marine nature notes: Cockles. *Otago Daily Times*, Issue 21609, 2 April 1932, 2.

——. (1932). Marine nature notes: Flounders. *Otago Daily Times*, Issue 21598, 19 March 1932, 2.

——. (1932). Marine nature notes: Groper. *Otago Daily Times*, Issue 21704, 23 July 1932, 2.

——. (1932). Marine nature notes: Hagfish or blind eel. *Otago Daily Times*, Issue 21586, 5 March 1932, 19.

——. (1932). Marine nature notes: John Dory. *Otago Daily Times*, Issue 21752, 17 September 1932, 2.

——. (1932). Marine nature notes: Smoothhound dogfish. *Otago Daily Times*, Issue 21656, 28 May 1932, 19.

——. (1933). Marine nature notes: New Zealand eels. *Otago Daily Times*, Issue 22149, 30 December 1933, 3.

—— (1933). Marine nature notes: Crabs. *Otago Daily Times*, Issue 21858, 21 January 1933, 19.

——. (1933). Marine nature notes: Frost fish. *Otago Daily Times*, Issue 21976, 10 June 1933, 19.

——. (1933). Marine nature notes: Kahawai. *Otago Daily Times*, Issue 21906, 18 March 1933, 20.

——. (1933). Marine nature notes: Local names of fish. *Otago Daily Times*, Issue 21952, 13 May 1933, 2.

——. (1933). Marine nature notes: Red gurnard. *Otago Daily Times*, Issue 21870, 4 February 1933, 2.

——. (1933). Marine nature notes: Rough leather jacket. *Otago Daily Times*, Issue 22060, 16 September 1933, 17.

——. (1934). Marine nature notes: Kelpfish. *Otago Daily Times*, Issue 22184, 10 February 1934, 2.

——. (1934). Marine nature notes: New Zealand eels. *Otago Daily Times*, Issue 22160, 13 January 1934, 18.

——. (1934). Marine nature notes: The kelp fish. *Otago Daily Times*, Issue 22172, 27 January 1934, 21.

——. (1935). Marine nature notes: Sea dogs. *Otago Daily Times*, Issue 22859, 18 April 1936, 5.

——. (1936). Marine nature notes: The kokopu or mountain trout. *Otago Daily Times*, Issue 22984, 12 September 1936, 5.

——. (1937). Marine nature notes: Freshwater crayfish. *Otago Daily Times*, Issue 23090, 16 January 1937, 21.

——. (1938). Fight against prejudice. *The Press*, Vol. LXXIV, Issue 22437, 25 June 1938, 19.

——. (1953). *A Treasury of New Zealand fish*. A. H. & A. W. Reed.

Grayling. [Exchange of fish.] (1895). *The Star*, Issue 5321, 27 July 1895, 6.

Greenhill, S. J., & Clark, R. (2011). POLLEX-Online: The Polynesian Lexicon Project Online. *Oceanic Linguistics*, 50(2), 551–559.

Grey, G. (1857). *Ko nga whakapepeha me nga whakaahuareka a nga tipuna o Aotea-roa*. Saul Solomon & Company.

——. (1885). *Polynesian mythology, and ancient traditional history of the Maori as told by their priests and chiefs* (No. 1). Whitcombe & Tombs.

Grey, Z. (1926). *Tales of the angler's Eldorado: New Zealand*. Rowman & Littlefield.

Haddon, M. (1994). Size-fecundity relationships, mating behaviour, and larval release in the New Zealand paddle crab, *Ovalipes catharus* (White 1843) (Brachyura: Portunidae). *New Zealand Journal of Marine and Freshwater Research*, 28(4), 329–334.

Hagfish bane of fishermen. (1948). *Bay of Plenty Times*, Vol. LXXVII, Issue 14821, 8 November 1948, 4.

Hamilton, A. (1908). *Fishing and sea-foods of the ancient Maori*. (No. 2). Government Printer.

Hardy, W. (1966). *The saltwater angler*. Murray.

Harvest of fish. (1931). *Thames Star*, Vol. LXV, Issue 18194, 20 May 1931, 4.

Heaphy, C. (1862). *A visit to the greenstone country*. G. T. Chapman.

Hector, J. (1872). *Fishes of New Zealand: Notes on the edible fishes*. James Hughes.

Hikuroa, D. (2017). Mātauranga Māori: The ūkaipō of knowledge in New Zealand. *Journal of the Royal Society of New Zealand*, 47(1), 5–10.

Hindmarsh, G., & Walker, M. (2015). *Kahawai: The people's fish*. Potton & Burton.

Hinojosa, I. A., Green, B. S., Gardner, C., Hesse, J., Stanley, J. A., & Jeffs, A. G. (2016). Reef sound as an orientation cue for shoreward migration by pueruli of the rock lobster, *Jasus edwardsii*. *PLoS One*, 11(6), e0157862.

Hodkinson, E. (1903). The paper nautilus. *New Zealand Illustrated Magazine*, Vol. VIII, Issue 1, 1 April 1903, 38.

Holder, C. F. (1897). A companion of the sunfish. *Lyttelton Times*, Vol. XCVII, Issue 11193, 15 February 1897, 2.

Hollows, J. (2016). *Freshwater crayfish farming: A guide to getting started*. Ernslaw One Ltd.

Hugo, V. (2010). *Toilers of the sea*. Project Gutenberg Australia. Retrieved from: www.gutenberg.org/ebooks/32338

Hursthouse, C. F. (1857). *New Zealand: Or, Zealandia, the Britain of the south (Vol. 2)*. E. Stanford. Retrieved from: Early New Zealand Books, University of Auckland.

Hutching, Gerard. 'Sharks and rays', Te Ara – the Encyclopedia of New Zealand. Retrieved from: www.TeAra.govt.nz/en/sharks-and-rays

J. C. (1927). About the koura. *Auckland Star*, Vol. LVIII, Issue 208, 3 September 1927, 21.

James, A. (2008). *Ecology of the New Zealand lamprey* (Geotria australis): *A literature review*. Department of Conservation.

Jellyman, D. (2014). Freshwater eels and people in New Zealand: A love/hate relationship. In Katsumi Tsukamoto & Mari Kuroki (eds), *Eels and humans*, 143–153. Springer.

Johnson, D., & Haworth, J. (2004). *Hooked: The story of the New Zealand fishing industry*. Hazard Press for the Fishing Industry Association.

Johnson, J. (2011). *The nutritional ecology of the New Zealand butterfish Odax pullus*. [Unpublished doctoral dissertation.] University of Auckland.

Jøn, A. A., & Aich, R. S. (2015). Southern shark lore forty years after *Jaws*: The positioning of sharks within Murihiku, New Zealand. *Australian Folklore*, 30.

Jones, H. F. E. (2011). *The ecological role of the suspension feeding bivalve, Austrovenus stutchburyi, in estuarine ecosystems.* [Unpublished doctoral dissertation.] University of Waikato.

Kapai Te Toheroa. (1926). *Rodney and Otamatea Times, Waitemata and Kaipara Gazette*, 1 September 1926, 5.

Keane, Basil. 'Taniwha – Sharks', Te Ara – *Encyclopedia of New Zealand*. Retrieved from: www.TeAra.govt.nz/en/taniwha/page-5

Keith, L. (1985). The birds of Waimea Inlet. *Journal of the Ornithological Society of New Zealand*, 271.

Kerekere, W. (1978). *Taupo Moana – Episode 1. The legend of the lake.* [Radio broadcast.] Ngā Taonga Sound & Vision, Reference number: 40489.

King, D., Tawhai, W., Skipper, A., & Iti, W. (2005). *Anticipating local weather and climate outcomes using Māori environmental indicators*. National Institute of Water & Atmospheric Research.

King, M. (2017). *Moriori: A people rediscovered*. Penguin Random House New Zealand.

Kingston, P. (Host). (1974). *Spectrum 114: No wonder they're a delicacy*. [Radio broadcast.] Ngā Taonga Sound & Vision, reference number: 30147.

Knapp, L., Mincarone, M. M., Harwell, H., Polidoro, B., Sanciangco, J., & Carpenter, K. (2011). Conservation status of the world's hagfish species and the loss of phylogenetic diversity and ecosystem function. *Aquatic Conservation: Marine and Freshwater Ecosystems*, 21(5), 401–411.

Knight, C. H. (2016). *New Zealand's rivers: An environmental history*. Canterbury University Press.

Kohere, R. T. (1951). *The autobiography of a Maori*. Reed.

Kusabs, I. A., Hicks, B. J., Quinn, J. M., & Hamilton, D. P. (2015). Sustainable management of freshwater crayfish (kōura, *Paranephrops planifrons*) in Te Arawa (Rotorua) lakes, North Island, New Zealand. *Fisheries Research*, 168, 35–46.

Kusabs, I. A., Hicks, B. J., Quinn, J. M., Perry, W. L., & Whaanga, H. (2018). Evaluation of a traditional Māori harvesting method for sampling kōura (freshwater crayfish, *Paranephrops planifrons*) and toi toi (bully, *Gobiomorphus* spp.) populations in two New Zealand streams. *New Zealand Journal of Marine and Freshwater Research*, 52(4), 603–625.

Land of possibilities: The future of toheroa. (1934). *Auckland Star*, Vol. LIV, Issue 99, 27 April 1923, 1.

Latter, B. (1940). Duvauchelle, head of the bay, *The Press*, Vol. LXXVI, Issue 23053, 22 June 1940, 15.

Leach, F., Davidson, J., Fraser, K., & Anderson, A. (1999). Pre-European catches of barracouta, *Thyrsites atun*, at Long Beach and Shag River Mouth, Otago, New Zealand. *Archaeofauna*, (8).

Lecture on acclimatisation. (1864). *Lyttelton Times*, Vol. XXI, Issue 1216, 31 March 1864, 5.

Lee, M. (2018). *Navigators & naturalists: French exploration of New Zealand and the Pacific (1769–1824)*. Bateman Books.

Leung, T. L. F., & Poulin, R. (2007). Interactions between parasites of the cockle *Austrovenus stutchburyi*: Hitch-hikers, resident-cleaners, and habitat-facilitators. *Parasitology*, 134(2), 247–255.

Local & general news. (1885). *Feilding Star*, Vol. VI, Issue 109, 28 February 1885, 2.

Lyric spiral, lighted whorl. (1948). *New Zealand Journal of Agriculture*, Vol. 76, Issue 2, 16 February 1948, 192.

McCarthy, A., Hepburn, C., Scott, N., Schweikert, K., Turner, R., & Moller, H. (2014). Local people see and care most? Severe depletion of inshore fisheries and its consequences for Māori communities in New Zealand. *Aquatic Conservation: Marine and Freshwater Ecosystems*, 24(3), 369–390.

McCaskill, L. W. (1937). Nature notes: Freshwater crayfish. *The Press*, Vol. LXXIII, Issue 22253, 18 November 1937, 6 (supplement).

——. (1938). Nature notes: The paua. *The Press*, Vol. LXXIV, Issue 22567, 24 November 1938, 5 (supplement).

McClain, C. R., Balk, M. A., Benfield, M. C., Branch, T. A., Chen, C., Cosgrove, J., Dove, A. D. M., Gaskins, L., Helm, R. R., Hochberg, F. G., Lee, F. B., Marhsall, A., McMurray, S. E., Schanche, C., Stone, S. N., & Thaler, A. D. (2015). Sizing ocean giants: patterns of intraspecific size variation in marine megafauna. *PeerJ*, 3, e715.

McCormack, G. (2007). Cook Islands biodiversity database, Version 2007.2. Cook Islands Natural Heritage Trust, Rarotonga. Retrieved from: http://cookislands.bishopmuseum.org

McDowall, R. M. (2003). Impacts of introduced salmonids on native galaxiids in New Zealand upland streams: A new look at an old problem. *Transactions of the American Fisheries Society*, 132(2), 229–238.

——. (2005). Historical biogeography of the New Zealand freshwater crayfishes (Parastacidae, *Paranephrops* spp.): Restoration of a refugial survivor? *New Zealand Journal of Zoology*, 32(2), 55–77.

——. (2010). *New Zealand freshwater fishes: An historical and ecological biogeography* (Vol. 32). Springer Science & Business Media.

——. (2011a). *Ikawai: Freshwater fishes in Māori culture and economy*. University of Canterbury.

——. (2011b). *Essays of a fishery scientist: 50 years of experience*. NIWA Information Series No. 80.

McEwan, A. J., Dobson-Waitere, A. R., & Shima, J. S. (2020). Comparing traditional and modern methods of kākahi translocation: Implications for ecological restoration. *New Zealand Journal of Marine and Freshwater Research*, 54(1), 102–114.

McIntosh, A. R., McHugh, P. A., Dunn, N. R., Goodman, J. M., Howard, S. W., Jellyman, P. G., O'Brien, L. K., Nyström, P., & Woodford, D. J. (2010). The impact of trout on galaxiid fishes in New Zealand. *New Zealand Journal of Ecology*, 34(1), 195–206.

McNab, R. (ed.). (1914). *Historical records of New Zealand (Vol. 2)*. Government Printer. Retrieved from: Early New Zealand Books, University of Auckland.

Mair, G. (1923). *Reminiscences and Maori stories*. Brett Printing and Publishing.

Malcolm, J. (1928). Food values of New Zealand fish. Part 9: Tinned toheroa and toheroa soup. *Transactions and Proceedings of the Royal Society of New Zealand*, Vol. 59, 1928, 85.

Maori exhibits. (1929). *Auckland Star*, Vol. LX, Issue 179, 31 July 1929, 10.

Mariner, G. R. (1908). The museum. *Wanganui Chronicle*, Vol. L, Issue 12145, 18 June 1908, 2.

Masterton. (1888). *New Zealand Times*, Vol. L, Issue 8325, 2 March 1888, 7.

Matthews, P. (1939). Flying fish for tea? *Auckland Star*, Vol. LXX, Issue 201, 26 August 1939, 16 (supplement).

Matthews, R. H. (1910). Reminiscences of Maori life fifty years ago. *Transactions and Proceedings of the Royal Society of New Zealand*, Vol. 43, 1910, 598.

Maxwell, K. H., Ngāti Horomoana, T. W. A. H., Arnold, R., & Dunn, M. R. (2018). Fishing for the cultural value of kahawai (*Arripis trutta*) at the Mōtū River, New Zealand. *New Zealand Journal of Marine and Freshwater Research*, 52(4), 557–576.

Mead, H. M., & Grove, N. (2004). *Nga pepeha a nga tipuna: The sayings of the ancestors*. Victoria University Press.

Melchior, M. (2021). Partitioning along reproductive niche dimensions in sympatric New Zealand freshwater bivalve species. [Unpublished doctoral dissertation.] University of Waikato.

Melchior, M., Squires, N. J., Clearwater, S. J., & Collier, K. J. (2021). Discovery of a host fish species for the threatened New Zealand freshwater mussel *Echyridella aucklandica* (Bivalvia: Unionida: Hyriidae). *New Zealand Journal of Marine and Freshwater Research*. Retrieved from: www.tandfonline.com/doi/full/10.1080/00288330.2021.1963290

Merrett, Joseph Jenner. (1844). *Maori feast at Remuera, 1844*. Star Steam Printing Company. Auckland Council Libraries. 995.1105 R38 (1840–49).

Mikaere, B. (1983). The obsidian island. *New Zealand Geographic*, Issue 003, July–September 1989. Retrieved from: www.nzgeo.com

Mitchell, S. J. (1984). Feeding of ling *Genypterus blacodes* (Bloch & Schneider) from four New Zealand offshore fishing grounds. *New Zealand Journal of Marine and Freshwater Research*, 18(3), 265–274.

Monk, A. (2017.) A growing tribal economy. *Te Karaka*, 76, 44–46.

Montgomery, J. C., & Saunders, A. J. (1985). Functional morphology of the piper *Hyporhamphus ihi* with reference to the role of the lateral line in feeding. *Proceedings of the Royal Society B: Biological Sciences*, 224, 197.

Moon, H. (1857). *An account of the wreck of H.M. sloop 'Osprey'*. Annett and Robinson.

Morrell, John. (2014). *Blackfish: A fisherman's journey*. [Documentary film.]

Morrison, M. A., Lowe, M. L., Jones, E. G., Makey, L., Shankar, U., Usmar, N., Miller, A., Smith, M., and Middleton, C. (2014). *Habitats of particular significance for fisheries management: The Kaipara Harbour.* New Zealand Aquatic Environment and Biodiversity Report No. 129. Ministry for Primary Industries.

Morton, J., & Miller, M. (1968). *The New Zealand sea shore.* HarperCollins.

Mossman, S. (2008). *Snapper: New Zealand's greatest fish.* AUT Media.

Mouritsen, K. N., & Poulin, R. (2003). Parasite-induced trophic facilitation exploited by a non-host predator: A manipulator's nightmare. *International Journal for Parasitology*, 33(10), 1043–1050.

——. (2003). The mud flat anemone-cockle association: Mutualism in the intertidal zone? *Oecologia*, 135(1), 131–137.

Murton, B. (2006). 'Toheroa Wars': Cultural politics and everyday resistance on a northern New Zealand beach. *New Zealand Geographer*, 62(1), 25–38.

Mysterious visitors. (1918). *The Press*, Vol. LIV, Issue 16286, 9 August 1918, 5.

New Zealand Department of Health. (1943). The problem of goitre. *New Zealand Journal of Agriculture*, Vol. 66, Issue 3, 15 March 1943, 180.

New Zealand eels: Their toll on trout. (1940). *Auckland Star*, Vol. LXXI, Issue 293, 10 December 1940, 5.

Nihoniho, T. (1913). *Narrative of the fighting on the East Coast 1865–71: With a monograph on bush fighting.* Dominion Museum.

Norman, M. D., Paul, D., Finn, J., & Tregenza, T. (2002). First encounter with a live male blanket octopus: The world's most sexually size-dimorphic large animal. *New Zealand Journal of Marine and Freshwater Research*, 36, 733–736.

Nyegaard, M., Sawai, E., Gemmell, N., Gillum, J., Loneragan, N. R., Yamanoue, Y., & Stewart, A. L. (2018). Hiding in broad daylight: Molecular and morphological data reveal a new ocean sunfish species (Tetraodontiformes: Molidae) that has eluded recognition. *Zoological Journal of the Linnean Society*, 182(3), 631–658.

O'Shea, S. (1999). The marine fauna of New Zealand: Octopoda (Mollusca: Cephalopoda). *NIWA Biodiversity Memoirs*, (112), 5–278.

Old-time frost-fishing. (1930). *Franklin Times*, Vol. XX, Issue 82, 18 July 1930, 8.

Orbell, M. (1985). *The natural world of the Maori.* Collins.

——. (1995). *The illustrated encyclopedia of Māori myth and legend.* Canterbury University Press.

Owen, A. (1969). *The whitebaiters.* [Radio documentary.] Ngā Taonga Sound & Vision, reference number: 331067.

Owen, K. L., & Sell, M. G. (1985). The birds of Waimea inlet. *Notornis*, 32, 271–309.

Pacoureau, N., Rigby, C. L., Kyne, P. M., Sherley, R. B., Winker, H., Carlson, J. K., Fordham, S. V., Barreto, R., Fernando, D., Francis, M. P., Jabado, R. W., Herman, K. B., Liu, K., Marshall, A. D., Pollom, R. A., Romanov, E. V., Simpfendorfer, C. A., Yin, J. S., Kindsvater, H. K., & Dulvy, N. K. (2021). Half a century of global decline in oceanic sharks and rays. *Nature*, 589(7843), 567–571.

Papers relating to special settlement on Stewart's Island. (1872). *Appendix to the Journals of the House of Representatives*, 1872 Session I, D-07a. Retrieved from: https://atojs.natlib.govt.nz

Parkyn, S., & Kusabs, I. (2007). *Taonga and mahinga kai species of the Te Arawa lakes: A review of current knowledge – kōura.* NIWA Client Report: HAM2007-022. National Institute of Water and Atmospheric Research.

Parliamentary Commissioner for the Environment. (2013). *On a pathway to extinction? An investigation into the status and management of the longfin eel*. Parliamentary Commissioner for the Environment: Wellington.

Parrott, A. W. (1957). *Sea angler's fishes of New Zealand*. Hodder & Stoughton.

——. (1958). *Big game fishes and sharks of New Zealand*. Hodder & Stoughton.

——. (1960). *The queer and the rare fishes of New Zealand*. Hodder & Stoughton.

Paul-Burke, K., Burke, J., Te Ūpokorehe Resource Management Team, Bluett, C., & Senior, T. (2018). Using Māori knowledge to assist understandings and management of shellfish populations in Ōhiwa harbour, Aotearoa New Zealand. *New Zealand Journal of Marine and Freshwater Research*, 52(4), 542–556.

Paulin, C. D. (2007). Perspectives of Māori fishing history and techniques. Ngā āhua me ngā pūrākau me ngā hangarau ika o te Māori. *Tuhinga: Records of the Museum of New Zealand Te Papa Tongarewa*, 18, 11–47.

Paulin, C., & Fenwick, M. (2016). *Te matau a Māui: Fish-hooks, fishing and fisheries in New Zealand*. University of Hawai'i Press.

Paulin, R. (1889). *The wild west coast of New Zealand: A summer cruise in the 'Rosa'*. Thoburn & Company.

Parkinson, S. (1984). *A journal of a voyage to the South Seas*. Caliban Books. Retrieved from: Early New Zealand Books, University of Auckland.

Penguin's fate: Swallowed by groper. (1938). *Thames Star*, Vol. LXVI, Issue 20229, 4 February 1938, 4.

Phillipps, W. J. (1947). A list of Maori fish names. *The Journal of the Polynesian Society*, 56(1), 41–51.

Polack, J. S. (1838). *New Zealand: Being a narrative of travels and adventures during a residence in that country between the years 1831 and 1837 (Vol. 1)*. Capper Reprint, 1974. Retrieved from: Early New Zealand Books, University of Auckland.

——. (1840). *Manners and customs of the New Zealanders: With notes corroborative of their habits, usages, etc., and remarks to intending emigrants, with numerous cuts drawn on wood (Vol. 1)*. J. Madden & Company.

Pollard, J. (1973) *Australian and New Zealand fishing*. Paul Hamlyn.

Pomare, M., & Cowan, J. (1977). *Legends of the Maori (Vol. 1)*. AMS Press.

Pond, W. (1997). *The land with all woods and waters*. Waitangi Tribunal. Retrieved from: www.waitangitribunal.govt.nz

Pope, A. (1829). *An essay on man: And other poems*. (Vol. 2). John Sharpe, Duke Street, Piccadilly.

Pope, E. C., Hays, G. C., Thys, T. M., Doyle, T. K., Sims, D. W., Queiroz, N., Hobson, V. J., Kubicek, L., & Houghton, J. D. (2010). The biology and ecology of the ocean sunfish Mola mola: A review of current knowledge and future research perspectives. *Reviews in Fish Biology and Fisheries*, 20(4), 471–487.

Powell, A. W. B. (1935). Nature notes: Our edible shellfish. *New Zealand Herald*, Vol. LXXII, Issue 22262, 9 November 1935, 1 (supplement).

——. (1937). Ways of the wild: The blue cod. *Auckland Star*, Vol. LXVIII, Issue 204, 28 August 1937, 1 (supplement).

——. (1941). Raise your hat mentally to the common shell. *Auckland Star*, Vol. LXXII, Issue 142, 18 June 1941, 6.

——. (1957). *Shells of New Zealand: An illustrated handbook*. Whitcombe & Tombs.

——. What is Australian salmon? *Auckland Star*, Vol. LXXII, Issue 103, 3 May 1941, 2 (supplement).

Pownall, Glen. (1971). *New Zealand shells and shell fish*. Seven Seas Publishing.

Radford, C. A., Ghazali, S. M., Montgomery, J. C., & Jeffs, A. G. (2016). Vocalisation repertoire of female bluefin gurnard (*Chelidonichthys kumu*) in captivity: sound structure, context and vocal activity. *PloS One*, 11(2), e0149338.

Radford, C. A., Putland, R. L., & Mensinger, A. F. (2018). Barking mad: The vocalisation of the John Dory, *Zeus faber. PloS One*, 13(10), e0204647.

Radford, C. A., Tay, K., & Goeritz, M. L. (2016). Hearing in the paddle crab, *Ovalipes catharus. Proceedings of Meetings on Acoustics*, 4ENAL, Vol. 27, 010013.

Radford, C., Jeffs, A., Tindle, C., & Montgomery, J. C. (2008). Resonating sea urchin skeletons create coastal choruses. *Marine Ecology Progress Series*, 362, 37–43.

Radway Allen, K. (1949). The New Zealand grayling – a vanishing species. *Tuatara*, Vol. 2, Issue 1, March 1949.

Raubenheimer, D., Zemke-White, W. L., Phillips, R. J., & Clements, K. D. (2005). Algal macronutrients and food selection by the omnivorous marine fish *Girella tricuspidata. Ecology*, 86(10), 2601–2610.

Recipes for preparing kina or sea eggs. (1965). *Te Ao Hou*, September 1965, 3.

Renata, A. (1892). Something more about salt water fish and fishing. *Otago Witness*, Issue 1994, 12 May 1892, 31.

——. (1911). The naturalist: By shore and pool. *Otago Witness*, Issue 2998, 30 August 1911, 76.

Riley, S. (Host). (1975). *Spectrum 158; Whitebait street.* [Radio broadcast.] Ngā Taonga Sound & Vision, reference number: 33205.

Rimini, T. W., Davies, G., & Tregear, E. (1901). Te puna kahawai i motu. *The Journal of the Polynesian Society*, 10(4), 183–190.

Roberts, C., Stewart, A. L., & Struthers, C. D. (eds). (2015). *The fishes of New Zealand*. Te Papa Press.

Roberts, M. (2013). Ways of seeing: Whakapapa. *Sites: A journal of social anthropology and cultural studies*, 10(1), 93–120.

Roberts, M., Norman, W., Minhinnick, N., Wihongi, D., & Kirkwood, C. (1995). Kaitiakitanga: Maori perspectives on conservation. *Pacific Conservation Biology*, 2(1), 7–20.

Roberts, M., Weko, F., & Clarke, L. (2006). *Maramataka: The Maori moon calendar*. Lincoln University Agribusiness and Economics Research Unit.

Robertson, D. A. (1980). Spawning of the frostfish, *Lepidopus caudatus* (Pisces: Trichiuridae), in New Zealand waters. *New Zealand Journal of Marine and Freshwater Research*, 14(2), 129–136.

Robson, C. H. (1875). Notes on the habits of the frost fish (*Lepidopus caudatus*). *Transactions and Proceedings of the Royal Society of New Zealand*, Vol. 8, 218.

Rod and line. (1938). *Waikato Times*, Vol. 122, Issue 20458, 26 March 1938, 22 (supplement).

Ross, P. M., Beentjes, M. P., Cope, J., De Lange, W. P., McFadgen, B. G., Redfearn, P., Searle, B., Skerrett, M., Smith, H., Smith, J., Te Tuhi, J., Tamihana, J., & Williams, J. R. (2018). The biology, ecology and history of toheroa (*Paphies ventricosa*): A review of scientific, local and customary knowledge. *New Zealand Journal of Marine and Freshwater Research*, 52(2), 196–231.

Ross, P. M., Knox, M. A., Smith, S., Smith, H., Williams, J., & Hogg, I. D. (2018). Historical translocations by Māori may explain the distribution and genetic structure of a threatened surf clam in Aotearoa (New Zealand). *Scientific Reports*, 8(1), 1–8.

Roughley, T. C. (1916). *Fishes of Australia and their technology (No. 21)*. W. A. Gullick, Government Printer.

Roy, E. A. (2020). 'Rolling and rolling and rolling': The first detailed account of great white shark sex. *The Guardian*, 4 April 2020. Retrieved from: www.theguardian.com

Ruddle, K. (1995). The role of validated local knowledge in the restoration of fisheries property rights: The example of the New Zealand Maori. *Property Rights in a Social and Ecological Context*, 2, 111–120.

Russell, B. C. (1983). The food and feeding habits of rocky reef fish of north-eastern New Zealand. *New Zealand Journal of Marine and Freshwater Research*, 17(2), 121–145.

Salmond, A. (1992). *Two worlds: First meetings between Maori and Europeans, 1642–1772*. University of Hawaii Press.

——. (1997). *Between worlds: Early exchanges between Maori and Europeans, 1773–1815*. Penguin Books (NZ) Ltd.

——. (2014). Tears of Rangi: Water, power, and people in New Zealand. *HAU: Journal of Ethnographic Theory*, 4(3), 285–309.

——. (2015). The fountain of fish: Ontological collisions at sea. In D. Bollier & S. Helfrich (eds), *Patterns of Commoning*, 309–329. The Commons Strategies Group.

Salmond, A., Brierley, G., & Hikuroa, D. (2019). Let the rivers speak: Thinking about waterways in Aotearoa New Zealand. *Policy Quarterly*, 15(3).

Samuel, E. (1936). The toheroa: New Zealand's exclusive shell-fish. *The New Zealand Railways Magazine*, Vol. 11, Issue 4.

Saunders, A. J., & Montgomery, J. C. (1985). Field and laboratory studies of the feeding behaviour of the piper Hyporhamphus ihi with reference to the role of the lateral line in feeding. *Proceedings of the Royal Society B: Biological Sciences*, 224, 209–221.

Savage, J. (1807). *Some account of New Zealand: Particularly the Bay of Islands, and surrounding country*. J. Murray. Retrieved from: Early New Zealand Books, University of Auckland.

Seized by an octopus. (1883). *Western Star*, Issue 704, 3 January 1883, 3.

Seized by an octopus. (1888). *Taranaki Herald*, Vol. XXXVII, Issue 8313, 6 November 1888, 2.

Sharman, A. R. (2019). *Mana wahine and atua wāhine*. [Unpublished master's thesis.] Victoria University of Wellington.

Shears, N. T., & Babcock, R. C. (2003). Continuing trophic cascade effects after 25 years of no-take marine reserve protection. *Marine Ecology Progress Series*, 246, 1–16.

Shellfish for Australia. (1936). *Stratford Evening Post*, Vol. IV, Issue 197, 31 July 1936, 5.

Sherrin, R. A. A. (1886). *Handbook of the fishes of New Zealand (Vol. 2)*. Wilson and Horton.

Shipping intelligence port of Oamaru. (1875). *North Otago Times*, Vol. XXI, Issue 1041, 13 March 1875, 2.

Shortland, T., & Tipene-Thomas, J. (2019) *Inventory of iwi and hapu eel research*. New Zealand Fisheries Assessment Report 2019/15.

Simpson, T. (1999). *A distant feast: The origins of New Zealand's cuisine*. Random House New Zealand.

Slaughter, R. J., Beasley, D. M. G., Lambie, B. S., & Schep, L. J. (2009). New Zealand's venomous creatures. *The New Zealand Medical Journal*, 122(1290), 83–97.

Some of our migratory fishes. (1890). *Ashburton Guardian*, Vol. VII, Issue 2361, 26 February 1890, 2.

Spyksma, A. J., Taylor, R. B., & Shears, N. T. (2017). Predation cues rather than resource availability promote cryptic behaviour in a habitat-forming sea urchin. *Oecologia*, 183(3), 821–829.

Stabbed by fish. (1938). *Manawatu Standard*, Vol. LIX, Issue 5, 2 December 1938, 12.

Stace, G. (1991). The elusive toheroa. *New Zealand Geographic*, 9, 18–34.

Stack, J. W. (1893). *Kaiapohia: The story of a siege*. Whitcombe & Tombs. Retrieved from: Early New Zealand Books, University of Auckland.

——. (1936). *More Maoriland adventures of J. W. Stack*. A. H. & A. W. Reed. Retrieved from: Early New Zealand Books, University of Auckland.

Stack, J. W. & E. (1938). *Further Maoriland adventures of J. W. and E. Stack – Book I. James West Stack's story*. A. H. & A. W. Reed. Retrieved from: Early New Zealand Books, University of Auckland.

Stanley, J. A., Radford, C. A., & Jeffs, A. G. (2012). Location, location, location: Finding a suitable home among the noise. *Proceedings of the Royal Society B: Biological Sciences*, 279(1742), 3622–3631.

Steel, F., Anderson, A., Ballantyne, T., Benjamin, J., Booth, D., Brickell, C., Gilderdale, P., Haines, D., Liebich, S., MacDiarmid, A., Maddison, B., McCarthy, A., Millar, G., Salesa, D., Scott, J., Stevens, M. J., & West, J. (2018). *New Zealand and the sea: Historical perspectives*. Bridget Williams Books.

Strange but true: Seven men and a sunfish. (1934). *North Canterbury Gazette*, Vol. 2, Issue 62, 16 March 1934, 5.

Swainson, W. (1859). *New Zealand and its colonization*. Smith, Elder. Retrieved from: Early New Zealand Books, University of Auckland.

Sykes, A., Chaney, J., & Delamere-Ririnui, K. (2020). *Brief of Evidence of Dr Joseph Selwyn Te Rito*. Annette Sykes & Co.

Symonds, J. E., Walker, S. P., Pether, S., Gublin, Y., McQueen, D., King, A., Irvine, G. W., Setiawan, A. N., Forsythe, J. A., & Bruce, M. (2014). Developing yellowtail kingfish (*Seriola lalandi*) and hāpuku (*Polyprion oxygeneios*) for New Zealand aquaculture. *New Zealand Journal of Marine and Freshwater Research*, 48(3), 371–384.

Tangiwai. (1931). Pictures of New Zealand life: Rotorua's winter. *The New Zealand Railways Magazine*, Vol. 6, Issue 3 (1 August 1931).

Taura, Y., van Schravendijk-Goodman, C., & Clarkson, B. (eds). (2017). *Te reo o te repo / The voice of the wetland: Connections, understandings and learnings for the restoration of our wetlands*. Manaaki Whenua – Landcare Research and Waikato Raupatu River Trust.

Taylor, D. I., & Schiel, D. R. (2010). Algal populations controlled by fish herbivory across a wave exposure gradient on southern temperate shores. *Ecology*, 91(1), 201–211.

Taylor, N. M. (1959). *Early travellers in New Zealand*. Oxford: Clarendon Press.

Taylor, R. (1855). *Te ika a Maui: Or, New Zealand and its inhabitants, illustrating the origin, manners, customs, mythology, religion, rites, songs, proverbs, fables, and language of the natives*. Cambridge University Press. Retrieved from: Early New Zealand Books, University of Auckland.

——. (1868). *The past and present of New Zealand: With its prospects for the future*. W. Macintosh. Retrieved from: Early New Zealand Books, University of Auckland.

Tedd, P. R., & Kelso, J. R. M. (1993). Distribution, growth and transformation timing of larval *Geotria australis* in New Zealand. *Ecology of Freshwater Fish*, 2(3), 99–107.

Te Rūnanga o Ngāi Tahu (Executive Producers). (2015). *Ngāi Tahu Mahinga Kai*. [Web series.] Te Rūnanga o Ngāi Tahu.

The blind eel. (1939). *Evening Star*, Issue 23270, 19 May 1939, 2.

The Church Missionary Society (1814–1851). Missionary Register. [Sections relating to New Zealand.] Retrieved from: Early New Zealand Books, University of Auckland.

The Evening Star. (1875). *Auckland Star*, Vol.VI, issue 17814, 15 November 1875, 2.

The Fisheries Court. (1890). *Otago Witness*, Issue 1980, 23 January 1890, 17.

The freshwater crayfish (after Leach). (1925). *The Star*, Issue 17580, 4 July 1925, 23 (supplement).

The lamprey. (1933). *Otago Daily Times*, Issue 22114, 18 November 1933, 9.

The paua: Its many uses. (1935). *Evening Post*, Vol. CXIX, Issue 128, 1 June 1935, 11.

The prince's chef. (1920). *Taihape Daily Times*, Vol. XI, Issue 3482, 10 May 1920, 8.

The sea fisheries of New Zealand. (1868). *Daily Southern Cross*, Vol. XXIV, Issue 3270, 9 January 1868, 4.

The story of Auckland. (1928). *Auckland Star*, Vol. LIX, Issue 209, 4 September 1928, 6.

The sun fish. (1896). *Otago Witness*, Issue 2206, 11 June 1896, 3 (supplement).

The visit of H.M.S. New Zealand. (1913). *Otago Daily Times*, Issue 15750, 29 April 1913, 4.

Thompson, C. (2019) *Sea people: The puzzle of Polynesia*. HarperCollins.

Thompson, J. T. (1874). The mythology and traditions of the Maori in New Zealand. *Transactions and Proceedings of the Royal Society of New Zealand*, Vol. 7.

Thompson, P. (1876). Fish and their seasons. *Transactions and Proceedings of the Royal Society of New Zealand*, Vol. 9.

Thys, T. M., Hays, G. C., & Houghton, J. D. (eds). (2020). *The ocean sunfishes: Evolution, biology and conservation*. CRC Press.

Toheroa popular: Milk bars in London. (1936). *Evening Post*, Vol. CXXI, Issue 144, 19 June 1936, 10.

Tolimieri, N., Jeffs, A., & Montgomery, J. C. (2000). Ambient sound as a cue for navigation by the pelagic larvae of reef fishes. *Marine Ecology Progress Series*, 207, 219–224.

Tragedy at coast. (1938). *Thames Star*, Vol. LXVI, Issue 20474, 28 November 1938, 2.

Travers, W. T. L. (1870). On the changes effected in the natural features of a new country by the introduction of civilized races. *Transactions and Proceedings of the Royal Society of New Zealand*, Vol. 2.

Travers, W. T. L., & Tikao, T. T. (1990). *Tikao talks: Ka taoka tapu o Te Ao Kohatu – Treasures from the ancient world of the Maori, told by Teone Taare Tikao to Herries Beattie*. Penguin Books.

Tregear, E. (1891). *Maori–Polynesian comparative dictionary*. Lyon and Blair.

——. (1904). *The Maori Race*. A. D. Willis. Retrieved from: New Zealand Electronic Text Collection, Victoria University of Wellington.

——. (1916). Maori mummies. *The Journal of the Polynesian Society*, Vol. 25, No. 100, 167–168.

Tremewan, C. (ed.). (2002). *Traditional stories from southern New Zealand*. Macmillan Brown Centre for Pacific Studies, University of Canterbury.

Trout fry liberated. (1926). *New Zealand Herald*, Vol. LXIII, Issue 19487, 17 November 1926, 13.

Tuna, the eel. (1911). *The Press*, Vol. LXVI, Issue 14233, 22 December 1911, 10.

Untitled. [Disposal of fish at Bluff.] (1911). *Southland Times*, Issue 16760, 13 June 1911, 4.

Untitled. [Record catch of greenbone.] (1929). *Otago Daily Times*, Issue 20629, 30 January 1929, 8.

Untitled. [Shag eaten by hapuku.] (1932). *Gisborne Times*, Vol. LXXIII, Issue 11616, 16 April 1932, 6.

Untitled. [Hapuku unable to be sold.] (1877). *Bruce Herald*, Vol. IX, Issue 875, 26 January 1877, 5.

Untitled. [Dynamiting grayling.] (1876). *Patea Mail*, Vol. II, Issue 141, 16 August 1876, 2.

Value of garfish. (1934). *Thames Star*, Vol. LXV, Issue 19205, 11 September 1934, 4.

Veart, D. (2013). *First, catch your weka: A story of New Zealand cooking*. Auckland University Press.

Visconti, V., Trip, E. D. L., Griffiths, M. H., & Clements, K. D. (2018). Life-history traits of the leatherjacket *Meuschenia scaber*, a long-lived monacanthid. *Journal of Fish Biology*, 92(2), 470–486.

——. (2018). Reproductive biology of the leatherjacket, *Meuschenia scaber* (Monacanthidae) (Forster 1801) in the Hauraki Gulf, New Zealand. *New Zealand Journal of Marine and Freshwater Research*, 52(1), 82–99.

——. (2020). Geographic variation in life-history traits of the long-lived monacanthid *Meuschenia scaber* (Monacanthidae). *Marine Biology*, 167(2), 1–13.

Visser, I. (1999). Benthic foraging on stingrays by killer whales (*Orcinus orca*) in New Zealand waters. *Marine Mammal Science*, 15(1), 220–227.

Waiapu fishing: Harvest of kahawai. (1947). *Gisborne Herald*, Vol. LXXIV, Issue 22248, 6 February 1947, 4.

Waitangi Tribunal. (1988). Wai 22 – *Report of the Waitangi Tribunal on the Muriwhenua Fishing Claim*. Retrieved from: www.waitangitribunal.govt.nz

——. (1992). Wai 27 – *The Ngai Tahu Sea Fisheries Report*. Retrieved from: www.waitangitribunal.govt.nz

——. (2001). Wai 64 – *REKOHU. A Report on Moriori and Ngati Mutunga Claims in the Chatham Islands*. Retrieved from: www.waitangitribunal.govt.nz

——. (2017). Wai 894 – Te Urewera. Vol. 1. Retrieved from: www.waitangitribunal.govt.nz

Wakefield, A. T., & Walker, L. (2005) *Maori methods and indicators for marine protection; Ngati Kere interests and expectations for the rohe moana*. Department of Conservation: Wellington.

Walsh, P. (1903). On the Maori method of preparing and using kokowai. *Transactions and Proceedings of the Royal Society of New Zealand*, Vol. 36, 4.

Ward, R. (1872). *Life among the Maories* [sic] *of New Zealand: Being a description of missionary, colonial, and military achievements*. G. Lamb.

Warne, K. (1989). Goat Island revisited. *New Zealand Geographic*, 01. Retrieved from: www.nzgeo.com

Wehi, P. M., and Roa, T. (2020). Reciprocal relationships: identity, tradition and food in the Kīngitanga Poukai He Manaakitanga: o te tuakiri, o te tikanga me te kai ki te Poukai o te Kīngitanga. *SocArXiv Papers*, Cornell University.

Wells, S. R. (2018). Changes to *Austrovenus stutchburyi* growth rate since early human settlement in New Zealand: An indication of the extent of human impact on estuarine health. [Unpublished doctoral dissertation.] University of Otago.

Whaanga, H., Wehi, P., Cox, M., Roa, T., & Kusabs, I. (2018). Māori oral traditions record and convey indigenous knowledge of marine and freshwater resources. *New Zealand Journal of Marine and Freshwater Research*, 52(4), 487–496.

White, J. (1885). *Maori customs and superstitions*. [Lectures from 1861.] Retrieved from: Early New Zealand Books, University of Auckland.

——. (1887). *The ancient history of the Maori, his mythology and traditions*. Cambridge University Press. Retrieved from: Early New Zealand Books, University of Auckland.

Williams, E., Crow, S., Murchie, A., Tipa, G., Egan, E., Kitson, J., Clearwater, S., & Fenwick, M. (2017). *Understanding taonga freshwater fish populations in Aotearoa-New Zealand. Prepared for Te Wai Māori Trust by the National Institute of Water and Atmospheric Research.* NIWA Client Report (2017326HN): Wellington, New Zealand.

Williams, W. L. (1892). On a specimen of sunfish captured at Poverty Bay. *Transactions and Proceedings of the Royal Society of New Zealand*, Vol. 25, 1892, 110.

Wolfe, K., Smith, A. M., Trimby, P., & Byrne, M. (2012). Vulnerability of the paper nautilus (*Argonauta nodosa*) shell to a climate-change ocean: Potential for extinction by dissolution. *The Biological Bulletin*, 223(2), 236–244.

Zintzen, V., Roberts, C. D., Anderson, M. J., Stewart, A. L., Struthers, C. D., & Harvey, E. S. (2011). Hagfish predatory behaviour and slime defence mechanism. *Scientific Reports*, 1(1), 1–6.

GLOSSARY

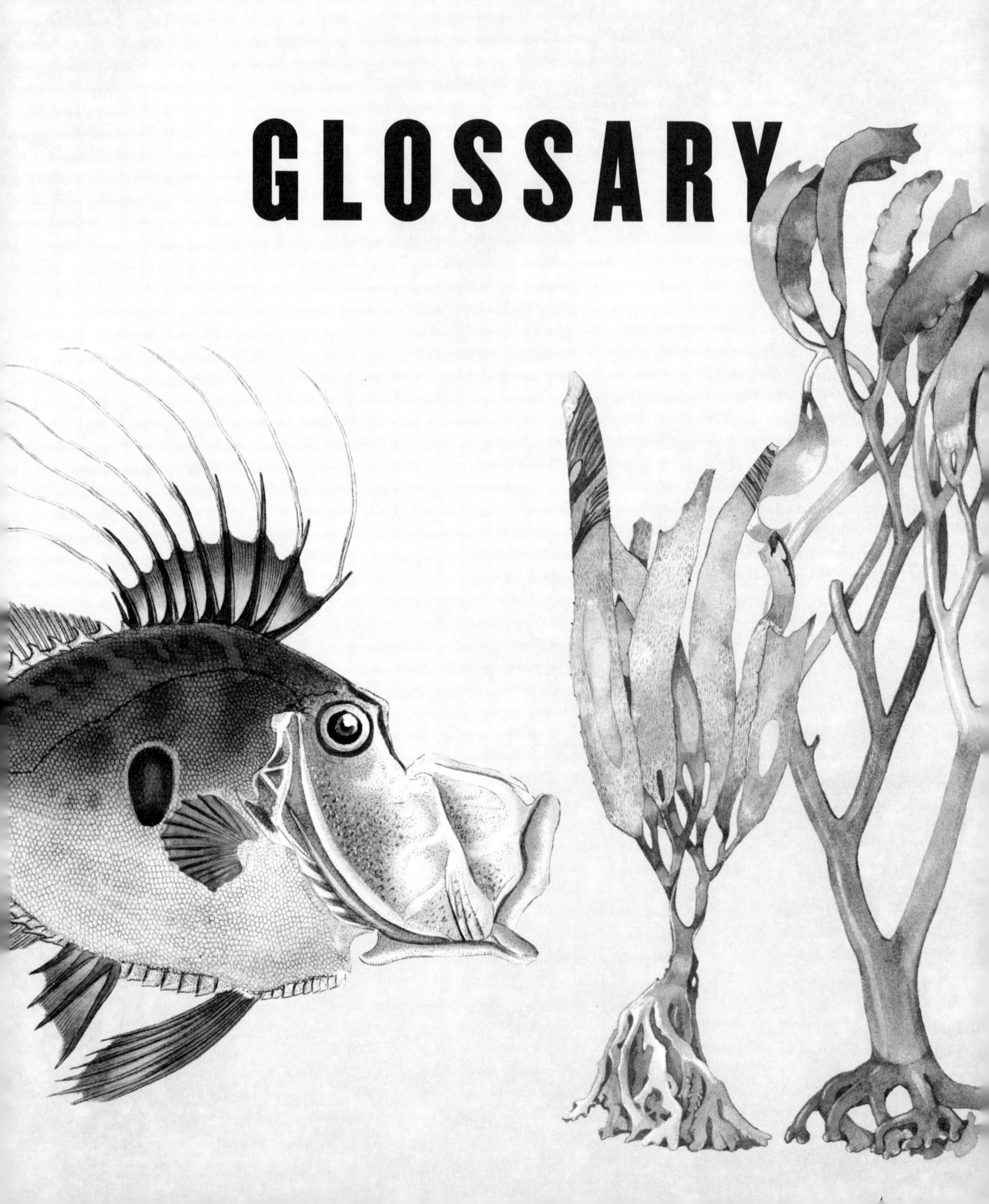

hapū kin group or sub-tribe
heke migration; also the name for migrating eels (tuna heke)
iwi tribe
kai food
kai moana seafood
kaihaukai exchange and sharing of food
kaitiaki guardian
kaniwha method of preparing raw fish for eating by immersing it in fresh water
karakia prayer, chant or incantation
kiekie a woody climbing vine, *Freycinetia banksii*
marae meeting house or meeting area
mana spiritual power, status, authority
māripi a knife or flat tool made of wood or bone, for removing pāua from rocks
matuku the Australasian bittern (*Botauris poiciloptilus*)
nene the base of the snapper's tongue
pā fortified village or settlement
pōhā a bag made of bull kelp seaweed, used for preserving food
pou carved wooden posts sometimes used to mark fishing grounds
pounamu greenstone
pūpū generic term for molluscs, shells, shellfish and snails
rangatira high-ranking chief or leader
rarawai an alternative name for hapuku, used while out at sea
rou kākahi a dredge for harvesting freshwater mussels
taonga treasure, or anything highly valued
tapu sacred or reserved from common use
tau kōura a method of collecting kōura using bundles of dried bracken fern (*Pteridium esculentum*)
Te Ao Māori the Māori world
Te Hiku o Te Ika 'the tail of the fish', a name for the Te Aupouri Peninsula and Cape Reinga
Te Ika a Māui 'the fish of Maui', an alternate name for the North Island of New Zealand
Te Tara o Te Ika 'the barb of the fish', a name for the Coromandel Peninsula
Te Upoko o Te Ika 'the head of the fish', a name for Wellington
tikanga customary protocols
tohunga a skilled expert
tūhua obsidian
urupā gravesite
whakairo wood carving
whakataukī proverb

DEITIES, LEGENDARY FIGURES AND PLACES

Hawaiki the spiritual homeland of the Māori people
Hina wife of Māui, and an atua associated with the moon
Hine-te-iwaiwa atua of childbirth and weaving
Māui a demigod and trickster responsible for many feats, including fishing up the North Island
Ngā rākau rua a Atuamatua the name of the voyaging waka that would later be named *Te Arawa*
puna-kauariki celestial waters in the sky
Tāne Mahuta atua of the forest and birds
Tangaroa atua of the oceans and sea creatures
taniwha a powerful creature that lives in water and takes on many forms
Tūmatauenga atua of war and the ancestor of humans
Tunaroa atua of eels
Te Parata a giant taniwha that creates the tides with its breath
Tāwhirimātea atua of the wind and storms
Tāwhaki demigod with powers of lightning
Tinirau a guardian of fish and a son of Tangaroa
Te Arawa original voyaging waka from which a number of iwi and hapū claim descent
tupua supernatural creatures similar to goblins

ACKNOWLEDGEMENTS

This book would not have been possible without the help of a huge number of people.

Thanks to the team at HarperCollins Publishers: Alex Hedley, Scott Forbes, Holly Hunter and a host of others, for taking the plunge with me on this project and for all of your expertise and hard work in making this book a reality.

A huge thanks to the talented Lars Quickfall, the 'Digital Da Vinci' from Boy Oh Boy productions, who was responsible for bringing the historic artworks back to life. A massive thanks also to the photographers Crispin and Irene Middleton from SeacologyNZ, Daan Hoffman, Ian Skipworth, Luke Colmer, Shaun Lee and many others who allowed me to use their beautiful underwater images. A huge thank you to my colleagues at Auckland War Memorial Museum – Tāmaki Paenga Hira for all their help and expertise, especially Wilma Blom, Tom Trnski, Severine Hannam, Zoe Richardson, Kelly Hall, Susan Tolich, Sarah Knowles and Rebecca Bray.

The content of this book is ultimately a compilation of the great work of others, and all credit is due to those authors for their work, while any mistakes or omissions are purely my own. A big thank you to the many researchers who took the time to answer my questions, in particular Marianne Nyegaard, Vanessa Taikato, Michele Melchior, Tom Moore, Craig Radford, John Montgomery, Christopher Scharpf and Richard Benton.

I relied extensively on a number of excellent databases and resources, namely the World Register of Marine Species (WORMS), The ETYfish Project, FishBase, POLLEX – Polynesian Lexicon Project Online, Alexander Turnbull Library, Biodiversity Heritage Library, The Early New Zealand Books Collection at the University of Auckland, The New Zealand Electronic Text Collection – Te Pūhikotuhi o Aotearoa, and Papers Past. David H. Graham's classic *A treasury of New Zealand fishes* and Rob McDowall's *Ikawai* were both inspirations for this book and I encourage anyone wanting to explore these topics further to seek them out.

Thanks to Colin Brickell for all the fish. Thanks to my brother, Chris, for helpful comments on the manuscript and thank you to Mum, Dad, Nana, Oketi, Kerry, and all of my family and friends for their support and encouragement. Finally, a huge thanks to my beautiful wife, Lizzy. I am truly grateful to spend my life with you and I love you with all of my heart.

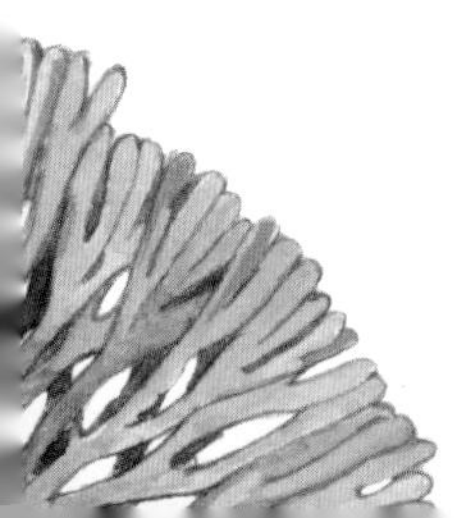

PICTURE CREDITS

Front cover
Top left: Sargassum sinclairii by Nancy Adams. (Te Papa, CA000892/001/0107)
Top right: Spotted eel (*Anguilla reinhardtii*) by Frank Olsen. (Department of Harbours and Marine, 1965; courtesy of Queensland Government)
Centre left: Gurnard (*Chelidonicthys kumu*) by Antoine Germain Bevalet, 1824. (Smithsonian Libraries)
Centre right: Nothogenia fastigiata by Nancy Adams. (Te Papa, CA000892/001/0049)
Centre right: Pāua (*Haliotis iris*) shell by Arthur Powell (1947), adapted by Lars Quickfall. (Auckland Museum)
Bottom left: Hormosira banksii by Nancy Adams. (Te Papa, CA000892/001/0084)
Bottom right: Schizoseris sp. by Nancy Adams. (Te Papa, CA000892/001/0062)

Back cover
Top left: Butterfish (*Odax pullus*) by Edgar Waite (1911), adapted by Lars Quickfall.
Top right: Pterocladia sp. by Nancy Adams. (Te Papa, CA000892/001/0042)
Centre left to right: Carpophyllum angustifolium by Nancy Adams. (Te Papa, CA000892/001/0096); *Pachymenia crassa* by Nancy Adams. (Te Papa, CA000892/001/0049); *Cystophora torulosa* by Nancy Adams. (Te Papa, CA000892/001/0097)
Bottom left: New Zealand cockles (*Austrovenus stutchburyi*) by Edgar Smith (c.1844), adapted by Lars Quickfall.

Page 1
Sargassum sinclairii by Nancy Adams. (Te Papa, CA000892/001/0107)
Carpophyllum angustifolium by Nancy Adams. (Te Papa, CA000892/001/0096)

Page 5
Butterfish (*Odax pullus*) by Edgar Waite, 1911. (Smithsonian Libraries)

Pages 6–7
Top right: Kahawai (*Arripis trutta*) by Frank Edward Clarke (1872) and Lars Quickfall. (Te Papa)
Left to right: Phycodrys quercifolia by Jean Baptiste Bory de Saint-Vincent, 1827. (Missouri Botanical Garden); *Hormosira banksii* by Nancy Adams. (Te Papa, CA000892/001/0084); *Cystophora torulosa* by Nancy Adams. (Te Papa, CA000892/001/0097)

Page 8
Left to right: Cystophora torulosa by Nancy Adams. (Te Papa, CA000892/001/0097); *Carpophyllum plumosum* by Nancy Adams. (Te Papa, CA000892/001/0092); *Hymenena durvillei* by Jean Baptiste Bory de Saint-Vincent, 1827. (Missouri Botanical Garden)

Page 10
Codium sp. by Nancy Adams. (Te Papa, CA000892/001/0101)

Page 12
Snapper (*Pagrus auratus*) by Alex Murray and Chas. Toms (1916), adapted by Lars Quickfall. (Auckland Museum)
Carpophyllum angustifolium by Nancy Adams. (Te Papa, CA000892/001/0096)

Pages 24–5
Top: Kōaro (*Galaxias brevipinnis*) by Frank Edward Clarke, c. 1887. (Te Papa, 1992-0035-2278/11)
Centre: Pouched lamprey (*Geotria australis*) by W. Wing (1851), adapted by Lars Quickfall. (University of Toronto)
Bottom: Kākahi (*Echyridella menziesii*) by Arthur Powell (1947), adapted by Lars Quickfall. (Auckland Museum)

Page 26–7
Top: Upokororo (*Prototroctes oxyrhynchus*) by Frank Edward Clarke, 1889. (Te Papa, 1992-0035-2278/1)
Left: Northern kōura (*Paranephrops planifrons*) by William Wing (c.1844), adapted by Lars Quickfall. (Biodiversity Heritage Library)
Right: Banded kōkopu (*Galaxias fasciatus*) by Joseph Drayton (1840). (Biodiversity Heritage Library)
Bottom: Giant kōkopu (*Galaxias grandis*) by James Francis McCardell, 1872. (Te Papa, MU000424/001/0043)

Pages 58–9
Top: Piper (*Hyporhamphus ihi*) by Frank Edward Clarke, 1875. (Te Papa, 1992-0035-2278/40)
Left: Short-tailed stingray (*Bathytoshia brevicaudata*) by A. R. Mc Culloch (1911), adapted by Lars Quickfall. (MBLWHOI Library)
Bottom: Toheroa (*Paphies ventricosa*) by Arthur Powell (1947), adapted by Lars Quickfall (Auckland Museum)

Pages 60–1
Top left: Sand flounder (*Rhombosolea plebeia*) by Frank Edward Clarke, 1873. (Te Papa, 1992-0035-2278/26)
Centre right: Paddle crab (*Ovalipes catharus*) by Kawahara Keiga (c. 1820s), adapted by Lars Quickfall. (Naturalis Biodiversity Center)
Top right: Gurnard (*Chelidonicthys kumu*) by Antoine Germain Bevalet, 1824. (Smithsonian Libraries)
Bottom: New Zealand cockles (*Austrovenus stutchburyi*) by Edgar Smith (c.1844), adapted by Lars Quickfall.
Bottom left: Codium sp. by Nancy Adams. (Te Papa, CA000892/001/0101)
Bottom right: Ulva sp. by Nancy Adams. Te Papa (CA000892/001/0001)

Pages 102–3
Top left: Snapper (*Pagrus auratus*) by Alex Murray and Chas. Toms (1916), adapted by Lars Quickfall.
Centre left: Male leatherjacket (*Meuschenia scaber*) by Edgar Waite (1911), adapted by Lars Quickfall.

Bottom left: Octopus illustration by Adolphe Millot (1910), adapted by Lars Quickfall.
Bottom centre: Kina (*Evechinus chloroticus*) by Arthur Powell (1947), adapted by Lars Quickfall.
Bottom right: Yellowfoot pāua (*Haliotis australis*) by H. Pilsbury (1890), adapted by Lars Quickfall.
Seaweeds:
Carpophyllum angustifolium and C. maschalocarpum by Nancy Adams. (Te Papa, CA000892/001/0096)

Pages 104–105
Top left: Female leatherjacket (*Meuschenia scaber*) by Edgar Waite (1911), adapted by Lars Quickfall.
Centre left: Parore (*Girella tricuspidata*) by Frank Edward Clarke (1896), adapted by Lars Quickfall.
Top right: Butterfish (*Odax pullus*) by Edgar Waite (1911), adapted by Lars Quickfall.

Seaweeds:
Carpophyllum angustifolium and C. maschalocarpum by Nancy Adams. (Te Papa, CA000892/001/0096)
Carpophyllum plumosum by Nancy Adams. (Te Papa, CA000892/001/0092)

Page 106
Carpophyllum angustifolium and C. maschalocarpum by Nancy Adams. (Te Papa, CA000892/001/0096)

Page 117
Lessonia variegata and *L. adamsiae* by Nancy Adams. (Te Papa, CA000892/002/0003)

Page 119
Left to right: Pachymenia crassa by Nancy Adams. (Te Papa, CA000892/001/0049) – also on p. 125; *Schizoseris* sp. by Nancy Adams. (Te Papa, CA000892/001/0062); *Nothogenia fastigiata* by Nancy Adams. (Te Papa, CA000892/001/0049)

Page 132
Hymenocladia sanguinea and *Cenacrum subsutum* by Nancy Adams. (Te Papa, CA000892/009/0002) – also on p. 137.
Iridaea lanceolata and *Iridea* sp. by Nancy Adams. (Te Papa, CA000892/009/0001)

Page 138
Ulva spp. by Nancy Adams. (Te Papa, CA000892/001/0001 & 0002) – also on p. 134.

Page 142
Carpophyllum angustifolium and C. maschalocarpum by Nancy Adams. (Te Papa, CA000892/001/0096) – also on p. 147.
Carpophyllum plumosum by Nancy Adams. (Te Papa, CA000892/001/0092) – also on p. 147.

Page 148
Macrocystis pyrifera by Nancy Adams. (Te Papa, CA000892/001/0108) – also on p. 153.
Macrocystis pyrifera by Jean Baptiste Bory de Saint-Vincent, 1827. (Missouri Botanical Garden) – also on p. 153.

Page 154
Lessonia variegata and *L. adamsiae* by Nancy Adams. (Te Papa, CA000892/002/0003)

Page 159
Lessonia variegata and *L. adamsiae* by Nancy Adams. (Te Papa, CA000892/002/0003)

Pages 160–1
Top left: Kingfish (*Seriola lalandi*) by Arthur Bartholomew (1885), adapted by Lars Quickfall. (Smithsonian Libraries)
Centre: Spiny dogfish (*Squalus acanthias*) by Frank Edward Clarke (c. 1870s–1880s).
Bottom: Great white shark (*Carcharodon carcharias*) by E. N. Fischer (1913), adapted by Lars Quickfall. (*Harvard University*)

Pages 162–3
Top left: Spotted dogfish (*Mustelus lenticulatus*) by Frank Olsen.
Top right: Kahawai (*Arripis trutta*) by Frank Edward Clarke (1872), adapted by Lars Quickfall. (Te Papa)
Centre right: Barracouta (*Thyrsites atun*) by William Below Gould (c.1832), adapted by Lars Quickfall.
Centre left: School shark (*Galeorhinus galeus*) by Frank Edward Clarke (c. 1870s–1880s).

Pages 198–9
Left: Ocean sunfish (*Mola mola*) by Kawahara Keiga, c.1820s. (Naturalis Biodiversity Center)
Right: Broad-gilled hagfish (*Eptatretus cirrhatus*) by Theodoor van Lith der jeude, c.1860s. (University of Amsterdam).

Pages 200–1
Frostfish (*Lepidopus caudatus*) by Jacques Reyne Isidore Acarie-Baron (c.1880s), adapted by Lars Quickfall. (University of Amsterdam)

Page 224
Ulva spp. by Nancy Adams. (Te Papa, CA000892/001/0001 & 0002)
Cystophora torulosa by Nancy Adams. (Te Papa, CA000892/001/0097)
Pāua (*Haliotis iris*) shell by Arthur Powell (1947), adapted by Lars Quickfall. (Auckland Museum)

Page 227
Caulerpa sertularioides by Nancy Adams. (Te Papa, CA000892/004/0001)

Page 231
Cystophora torulosa by Nancy Adams. (Te Papa, CA000892/001/0097)

Page 232
Hymenocladia sanguinea and *Cenacrum subsutum* by Nancy Adams. (Te Papa, CA000892/009/0002)
Iridaea lanceolata and *Iridea* sp. by Nancy Adams. (Te Papa, CA000892/009/0001)

Page 249
Pterocladia sp. By Nancy Adams. (Te Papa, CA000892/001/0042)

Page 250
John Dory (*Zeus faber*) by Edward Donovan, c.1808. (Biodiversity Heritage Library)
Lessonia variegata and *L. adamsiae* by Nancy Adams. (Te Papa, CA000892/002/0003)

Page 252
Left to right: Pachymenia crassa by Nancy Adams. (Te Papa, CA000892/001/0049); *Schizoseris* sp. by Nancy Adams. (Te Papa, CA000892/001/0062); *Nothogenia fastigiata* by Nancy Adams. (Te Papa, CA000892/001/0049)

Page 256
Sargassum sinclairii by Nancy Adams. (Te Papa, CA000892/001/0107)
Carpophyllum angustifolium by Nancy Adams. (Te Papa, CA000892/001/0096)